Praise for

Mozart's Starling

Selected as one of the Best Books of the Year by NPR

"Warmly underscoring the bonds between humans and animals, *Mozart's Starling* blends committed nature writing with personal reflection, music, and humor—and guarantees you will never look at the small, speckled birds in quite the same way again." —Tom Huizenga, National Public Radio

"This hard-to-put-down, charming blend of science, biography, and memoir illuminating the little-known story of the composer and his beloved bird is enlivened by the immediacy of Haupt's tales of her pet starling, Carmen, and brimming with starling information, travelogues, and historical details about Mozart's Vienna."

—Nancy Bent, *Booklist*

"Weaving together cheerful memoir, natural history, and biography, the author celebrates her 'insatiably social' pet starling, Carmen; investigates Mozart's experience with his avian companion . . . and offers intriguing details about starling behavior."

—*Kirkus Reviews*

"Delightful and interesting."

—Irene Wanner, *Seattle Times*

"Charming and highly readable."

—Sarah Bryan Miller, *St. Louis Post-Dispatch*

"An unusual and thoroughly engaging memoir."

—Sarah Murdoch, *Toronto Star*

"This book will surprise and delight you."

—Gene Walz, *Winnipeg Free Press*

"Readers will be led to contemplate the respective roles that humans and birds play on Earth. The book delves much deeper than birds and humans, though. It can be a humanities textbook that explores the blending of human culture and the natural world."

—David Hendricks, *San Francisco Chronicle*

"*Mozart's Starling* sparkles with imagination, emotion, and insight. Common birds, who too many consider vermin, have great gifts to share. Thank you, Lyanda Lynn Haupt, for showing us the delight and magic of a starling."

—Sy Montgomery, author of *Birdology* and *The Soul of an Octopus*

"By raising her own pet starling, Lyanda Lynn Haupt reveals something that music historians have missed—how daily life with a bird impacted Mozart during his most productive period. By sharing this delightful tale with the rest of us, she also reveals the unexpected quirks and charms of a species too often dismissed as a pest. *Mozart's Starling* is pure pleasure."

—Thor Hanson, author of *The Triumph of Seeds*

"*Mozart's Starling* is a delightful, enlightening, breathless flight through the worlds of Carmen and Star, two European starlings who join their human counterparts in exploring life and music and nature, helping to shed light on the connection between humans and birds—those of us bound to terra firma, and those who are free to soar."

—Garth Stein, author of *The Art of Racing in the Rain* and *A Sudden Light*

"I've long been a fan of Lyanda Lynn Haupt's writing, but in *Mozart's Starling* she wings it to another level. From the few but beguiling wisps that have come down about the pet starling that Mozart harbored for a couple of years, Haupt soars through a wide-ranging meditation on music, mimicry, language, Viennese manners and mores, avian behavior, perception of time and space, and the skein of spirit that connects humans to the creatures around them, including the much-reviled starling. The rescue and rearing of her own pet starling, Carmen, by turns harrowing and hilarious, is a deeply satisfying emotional counterpoint. I came away utterly convinced that Mozart was himself starling-like in his mischievous, quicksilver, sometimes raunchy, sometimes celestial genius. This volume sent me outside with a song in my heart and a glint in my eye as I surveyed the sky for the magic Haupt conjures up on every page."

—David Laskin, author of *The Children's Blizzard* and *The Family*

"Haupt's book entertainingly entwines two tales: what is both known and surmised about the life of Mozart and his pet starling, and the actual facts of life about living with such a creature. Both tales are engaging, and a naive reader will learn quite a bit, whether Haupt is discussing musical history, avian behavior, or the sometimes unexpected influences they have upon each other. Someone with an already strong love of Mozart and interest in birds will come away with an even deeper appreciation of both."

—Irene M. Pepperberg, PhD, research associate, Harvard University, and author of *Alex & Me*

ALSO BY Lyanda Lynn Haupt

The Urban Bestiary: Encountering the Everyday Wild

Crow Planet: Essential Wisdom from the Urban Wilderness

Pilgrim on the Great Bird Continent: The Importance of Everything and Other Lessons from Darwin's Lost Notebooks

Rare Encounters with Ordinary Birds

MOZART'S STARLING

LYANDA LYNN HAUPT

BACK BAY BOOKS
Little, Brown and Company
New York Boston Toronto

For Ginny

—who brings music to our lives

Back Bay Books / Little, Brown and Company
Hachette Book Group
1290 Avenue of the Americas, New York, NY 10104
littlebrown.com

Originally published in hardcover by Little, Brown and Company, April 2017
First Back Bay paperback edition, May 2018

Back Bay Books is an imprint of Little, Brown and Company,
a division of Hachette Book Group, Inc.
The Back Bay Books name and logo are trademarks of Hachette Book Group, Inc.

ISBN 978-0-316-37089-9 (hc) / 978-0-316-37090-5 (pb)
LCCN 2016954839

10 9 8 7 6 5 4 3 2 1

LSC-C

Printed in the United States of America

It does seem hard that our earth may be a far better place than we have yet discovered, and that peace and content may be only round the corner, yet that somehow our song of praise is prevented, or does not go well with Hesperus, unlike that of a silly bird.

—H. M. Tomlinson, *The Face of the Earth*

CONTENTS

MOZART'S STARLING

Prelude

A PLAGUE OF INSPIRATION

This book would have taken me half as long to write if it were not for one fact: most of it was composed with a starling perched on my shoulder. Or at least in the *vicinity* of my shoulder. Sometimes she was standing on top of my head. Sometimes she was nudging the tips of my fingers as they attempted to tap the computer keys. Sometimes she was defoliating the Post-it notes from books where I had carefully placed them to mark passages essential to the chapter I was working on—she would stand there in a cloud of tiny pink and yellow papers with an expression on her intelligent face that I could only read as *pleased*. She pooped on my screen. She pooped in my hair. Sometimes she would watch, with me, the chickadees that came to my window feeder to nibble the sunflower seeds I left for them. Sometimes she would look me in the eye and say, *Hi, honey!* Clear as day. "Hi, Carmen," I would whisper back to her.

Sometimes, tired of all these things and seemingly unable to come up with a new way to entertain herself or pester me, she would stand close to my neck, tunnel beneath my hair, and nestle down, covering her warm little feet with her soft breast feathers, so close to my ear that I could hear her heartbeat. She would close her eyes and fall into a light bird sleep.

It sounds like a sweet scene, but there is a conflict at its center. I am a nature writer, a birdwatcher, and a committed wildlife advocate, so the fact that I have lovingly raised a European starling in my living room is something of a confession. In conservation circles, starlings are easily the most despised birds in all of North America, and with good reason. They are a ubiquitous, nonnative, invasive species that feasts insatiably upon agricultural crops, invades sensitive habitats, outcompetes native birds for food and nest sites, and creates way too much poop. Millions of starlings have spread across the continent since they were introduced from England into New York's Central Park one hundred and thirty years ago.

An adult starling is about eight and a half inches from tip to tail, a fair bit larger than a sparrow but still smaller than a robin, with iridescent black feathers and a long, sharp, pointed bill. Just over a hundred and fifty years before the first starlings appeared in Central Park, the Swedish botanist and zoologist Carl Linnaeus had placed the species within his emerging avian taxonomy and christened it with the Latinized name we still use: *Sturnus vulgaris. Sturnus* for "star," referring to the shape of the bird in flight, with its pointed wings, bill, and tail; and *vulgaris,* not for "vulgar," as starling detractors like to assume, but for "common."* When Linnaeus named the bird, it was simply part of the European landscape and had not spread across the waters. There was no controversy surrounding the species; it was just a pretty bird. Starlings are now one of the most pervasive birds in North America, and there are so many that no one can count them; estimates run to about two hundred million. Ecologically, their presence here lies on a scale somewhere between highly unfortunate and utterly disastrous.

In *The Birdist's Rules of Birding,* a National Audubon Society blog by environmental journalist Nicholas Lund, one of the primary rules is actually "It's Okay to Hate Starlings." Sometimes beginning birders in the first flush of bird-love believe that it is a requirement of their newfound vocation to be enamored of all feathered creatures. But as we learn more, writes Lund, our relationships with various species become more nuanced. Some species are universally loved; who wouldn't feel happy in the presence of a cheerful black-capped chickadee? But once we become more informed about starlings, we begin to feel an inner dissonance. Lund tells birders who are first experiencing such confusion not to feel guilty: "It's okay to hate certain species . . . healthy, even. I suggest you start with European Starlings." In addition to

* Some suggest that the *star* part of the name refers to the little white spots that shimmer on the tips of the bird's black feathers during the nonbreeding season. It is impossible to know the genesis of the name for certain.

the issues with starlings I've listed, Lund adds: "They're loud and annoying, and they're everywhere."

It's true; among those who know a little about North American birds, starlings are not just disliked, they're outright hated. In *The Thing with Feathers: The Surprising Lives of Birds and What They Reveal About Being Human,* birder and journalist Noah Strycker (famous for seeing more species of birds on earth in one year than anyone, ever) writes, "If you Google 'America's most hated bird,' all of the top results refer to starlings. Such universal agreement is rare in matters of opinion, but on this everyone seems to concur: Starlings are rats with wings." Birders typically keep lists of the species they see on a field trip, but many don't even include the invasive starling on their tallies. Ornithological writer and blogger Chris Petrak does list them, not because he is glad to spot them but because he is "interested in those rare occasions when I can go almost an entire day without seeing a starling, and those even rarer days when I don't see one at all." The joy of a starling-less list. He goes on to back up Strycker: "Bird lover or not, the starling is not a loved bird. In fact, it is without a doubt the most hated bird in America."

Common, invasive, aggressive, reviled. Starlings don't just lie beneath our notice, the sentiment runs, they are actually *undeserving* of our notice. By rights, I know I should agree with the many guests in my home who learn that a starling lives here and pronounce, "Oh, I am a bird-lover, so I hate starlings." I *do* detest the presence of the species in North America. But this bird on my shoulder? Mischievous, clever, disorderly, pestering, sparkling, sleepy? Yes, I confess, I couldn't be more fond of her.

People always ask how I get the ideas for my books; I think all authors hear this question. And, at least for me, there is only one answer: You can't think up an idea. Instead, an idea flies into your brain—unbidden, careening, and wild, like a bird out of the ether. And though there is a measure of chance, luck, and grace involved, for the most part ideas don't rise from *actual* ether; instead, they spring from the metaphoric opposite—from the rich soil that has been prepared, with and without our knowledge, by the whole of our lives: what we do, what we know, what we see, what we dream, what we fear, what we love.

For much of my life I have studied birds. I have watched them, sketched them, scribbled notes about their habits and habitats. I have spent hundreds of hours immersed in ornithological texts and journals. I worked for a time as a raptor rehabilitator, and once I had done that, it seemed that all the injured birds within a fifty-mile radius had a way of finding me. People discovered wounded birds in their backyards and brought them to me in small boxes. A flopping, broken-legged gull turned up on my doorstep. One day while I was out for a walk, a diseased robin fell from a tree and landed on the sidewalk literally at my feet. And though I left rehab behind long ago, I have too often found myself raising orphaned chicks of various species, or binding the wings of injured birds, or making sick birds comfortable as they pass into the next world. So it makes sense that my thoughts, my life, and my work have been inspired by birds. But not by starlings. Because my subjects included everyday nature and urban wildlife, I had written about starlings out of necessity, but not out of true *inspiration.* Starlings, I felt, deserved no such esteem.

And as a writer, of course, I live by inspiration. I watch it come and go; when it's missing, I pray for its reappearance. I light a candle and put it in my window hoping that this little ritual might help inspiration find its way back to me, like a lover lost in a snowstorm. The word itself is beautiful. *Inspire* is from the Latin meaning "to be breathed upon; to be *breathed into.*" Just as I ponder the migrations of birds, I ponder the migrations of inspiration's light breeze. If it's not with me, where is it? Where has it been? Who has it breathed upon while it was away, and when, and how? Over and over again, I have come to terms with the sad truth that inspiration never visits at my convenience, nor in accordance with my sense of timing, nor at the behest of my will. Most of all, the inspiration-wind has no interest whatsoever in what I *think* I want to write about.

One day a couple of years ago I was gazing out of my study window and noticed a plague of starlings on the grassy parking strip in front of the house.* I was not looking for an idea that day—I had an engaging project on my desk and was just pondering the next sentence, not the next book. I pounded on the window to scare the starlings away, as I often do when they gather in numbers. The other little neighborhood birds find groups of starlings menacing—when starlings descend, the chickadees in my hawthorn tree rush away, as do the bushtits, and even the larger robins. Only the bold crows remain. So I pounded. The starlings flapped and rose halfheartedly, then landed again and returned to their grubbing for worms in the parking-strip grass. I rapped the window harder, and again they lifted. But this time, they turned toward the light and I was dazzled by the glistening iridescence of their breasts. So shimmery, ink black and scattered with pearlescent spots, like snow in sun. Hated birds, lovely birds. In this moment of conflicted beauty, a story I'd heard many times leapt to mind.

Mozart kept a pet starling. I can't even remember where I read that in my ornithological studies—it is one of those arcane little details recorded here and there, usually without substantiation. I repeated it myself in my first book, *Rare Encounters with Ordinary Birds.* Later, I was reading Jim Lynch's lovely novel *Border Songs* and discovered that one of his characters mentioned it. When I asked Lynch where he'd heard about Mozart's starling, he told me, "I read it in

* There are exaltations of larks and murders of crows. A flock of flying starlings is called, beautifully, a murmuration, but there is no official name for a terrestrial flock, as far as I know. *Plague* seems appropriate.

your book." Oh, dear! I began to worry that I'd been spreading an apocryphal story, but further research assured me that the tale was true. Mozart discovered the starling in a Vienna pet shop, where the bird had somehow learned to sing the motif from his newest piano concerto. Enchanted, he bought the bird for a few kreuzer and kept it for three years before it died. Just *how* the starling learned Mozart's motif is a wonderful musico-ornithological mystery. But there is one thing we know for certain: Mozart loved his starling. Recent examination of his work during and after the period he lived with the bird shows that the starling influenced his music and, I believe, at least one of the opera world's favorite characters. The starling was in turn his companion, distraction, consolation, and muse. When his father, Leopold, died, Wolfgang did not travel to Salzburg for the services. When his starling died, two months later, Mozart hosted a formal funeral in his garden and composed a whimsical elegy that proclaimed his affinity with the starling's friendly mischievousness and his sorrow over the bird's loss.

Mozart is only one of many composers and artists throughout the centuries who've had birds as pets. Mozart kept canaries, too, at different times in his life. But the fact that Mozart lived with, and loved, a *starling* is extraordinary. One of the world's greatest composers chose, as a household companion, what is now one of the world's most hated birds. I have spoken with classical music lovers who are offended at the very notion that Mozart might have been inspired by this invasive species, and birdwatchers are just as indignant. What good could be associated with a starling? Along with our understanding that starlings are common and unwelcome arises an assumption that we humans tend to attach to *all* things common and unwelcome: that they are also dirty, ugly, disease-ridden, and probably dumb—certainly not proper consorts for genius.

While I was looking out that day at the pearly-snow-breasted starlings, while I was thinking of their despisedness and their loveliness and Mozart in one swirl, I noticed the music pouring from my iPhone Pandora station. It was Mozart's Prague Symphony. Other than being composed by Mozart, this symphony has little to do with the tale of his pet bird (it was written while they lived together, though I didn't know this at the time). But the synchronicity was enough for me. The hair on the back of my neck prickled as I felt a new obsession take root in my psyche. I could not stop wondering over the tangled story of Mozart and his starling and felt that I was being pulled through an unseen gateway as I began to follow the tale's trail, uncovering all that I could from my two-hundred-and-fifty-year remove.

What did Mozart learn from his bird? The juxtaposition of the hated and sublime is fascinating enough. But how did they interact? What was the source of their affinity? And how did the starling come to know Mozart's tune? I dove into research, poring over the academic literature. I took to the streets, making detailed notes on the starlings in my neighborhood. But gaps in my understanding of starling behavior remained and niggled, and within a few weeks I reluctantly realized that to truly understand what it meant for Mozart to live with a starling, I would, like the maestro, have to live with a starling of my own.

I'd raised several starlings while working as a raptor rehabilitator for the Vermont Institute of Natural Science many years ago. Starlings aren't raptors, of course, but people brought us all kinds of birds. It was the official policy of the rehab facility to euthanize any starlings that came through the door rather than lavish scarce resources on them and then release them into the wild to wreak their ecological havoc. Most often the starlings that came to us were babies, orphaned or cat-caught; the people who brought them had no idea about the ecological conflict and usually didn't even know what kind of bird they had. They were just filled with compassion for another creature that needed care and had gone out of their way to act on their feelings as best they could. One little boy, about eight years old, carefully held out a baby starling cradled in a beautiful nest he'd made of grass and tissue. "Can you help him?" he asked with wide, expectant eyes as his mother stood watching behind him. What was I supposed to say? *Sure, honey. Give me the bird—I'll wring his scruffy neck for you.* It seemed to me that the lessons to be gleaned in terms of respect for life and compassion for other creatures outweighed any slight ecological impact the release of a few individual starlings might have.

So I became a renegade rehabber and made a deal with the folks who brought starlings in: I'd tend the chicks on my own time while they were in the precarious nestling phase, then give them back to their young rescuers for final raising and release.

It was fun to have juvenile starlings around the house; they were smart, busy, social, sweet, and made wonderful companions. But that was when I lived in a group house with a bunch of other hippie graduate-student ornithologists; having wild birds roaming around and a little bird poop here and there seemed perfectly normal. I brought all manner of birds home, from hummingbirds to hawks, and even great horned owls, which my housemates made me keep in the laundry room because they smelled of their last meal (skunk). And I'd always said good-bye to these starlings once they were minimally self-sufficient, not after they'd grown into aggressive, adult birds. What would it be like, I thought now, to raise a starling for months, maybe years, in my grown-up household where I had decent furniture, expensive musical instruments, work to accomplish, and guests who would think I was batshit crazy?

It turns out that one little bird was capable of turning my household, and my brain, completely upside down. I thought I was bringing a wild starling into my home as a form of research for this book, but this bird had ideas of her own. Instead of settling dutifully into her role as the subject of my grandiose social-scientific-musical experiment, Carmen turned the tables. She became the teacher, the guide, and I became an unwitting student—or, more accurately, a pilgrim, a wondering journeyer who had no idea what was to come. Following Mozart's starling, and mine, I was led on a crooked, beautiful, and unexpected path that wound through Vienna and Salzburg, the symphony, the opera, ornithological labs, the depths of music theory, and the field of linguistics. It led me to outer space. It led me deep into the spirit of the natural world and our constant wild animal companions. It led me to the understanding that there is more possibility in our relationships with animals—with all the creatures of the earth, not just the traditionally beautiful, or endangered, or loved—than I had ever imagined. And in this potential for relationship there lies our deepest source of creativity, of sustenance, of intelligence, and of *inspiration*. Before all of this, though, I learned that obtaining a starling, as abundant and legally unprotected as they may be, is not as easy as you'd think. Mozart paid a few coins for his bird in a shop. My route to acquiring a starling housemate was a bit more complicated.

One

THE STARLING OF SEATTLE

The details of Carmen's coming to live with me are admittedly a bit sketchy—part rescue, part theft. A friend turned informant (who prefers to remain anonymous—I'll call him Phil) knew I was on the lookout for an orphaned starling chick. He worked for the parks department and let me know that the starling nests under the restroom roof at a park near my home were slated for removal (or "sweeping"). I was aware of these nests and had been checking on their occupants' progress—the chirring sounds coming from beneath the eaves told me that the babies had already hatched. When I mentioned this to Phil, he said, "Yeah, well, you know they're just starlings." Park officials do attempt to remove the nests of unprotected pest species *before* chicks emerge from their eggs, but sometimes the timing doesn't work out, and the nests are removed anyway. It is illegal under the Migratory Bird Treaty Act to disturb, or even touch, the nests of most birds, but anyone—government official or private citizen—may with impunity destroy the nests and eggs of starlings and kill the nestlings and adult birds any way he or she likes. As nonnative invasive species, starlings, along with house sparrows and pigeons, have no legal standing or protection.*

When federal or state fish and wildlife departments do work that involves the killing of animals (like the shooting of overpopulous Canada geese or white-tailed deer, or the trapping or shooting of urban coyotes), it is usually accomplished under cover of darkness to prevent protests by well-intentioned animal lovers. Starling nest removal is no different. "It's going down tonight," Phil reported on the starling nest sweep, and I giggled to myself, suddenly feeling like I was part of a bank heist. I thanked Phil and arranged to meet my husband, Tom, at the park after work—I would need his help. While it is legal to pluck a baby starling from its nest, it would likely be misunderstood by any observers, and I didn't want to draw attention. The park was in high use that evening, with thirty little boys running around in soccer cleats as their coach yelled instructions in a lovely Welsh accent. We scoped out the nest that seemed easiest to access and nonchalantly dragged the giant plastic park garbage can into the men's

* Pigeons are officially considered feral rather than invasive. Early in this country's history, rock pigeons (the common urban species) were brought over from England, propagated, kept by settlers, and carried along on journeys west as sources of food. All the urban pigeons we see today are descendants of these pioneer pigeons, many of whom escaped. Their native habitat includes rocky cliffs, and we can imagine them in such places when we see them on high city buildings.

room. Tom climbed on top of it, slipped his long arm between the top of the wall and the eave of the roofline, and stretched toward the chirring sound. "Can't reach," he announced, withdrawing his arm, scraped from the effort. We switched places, thinking my smaller arm might slip through the wooden slats more easily, which it did. I stood on my toes, felt over the matted grassy nest stuffs, and stretched as far as I was able. I could actually feel the warmth radiating from the bodies of the nestlings, but while my arm was thinner than Tom's (muscles, he likes to point out), it was also shorter. I couldn't get any closer to the chicks, and I gained a deeper appreciation for the starling nesting strategy: they choose cavities that are set back far enough to be out of the reach of nest-thieving predators (more often a crow or a raccoon than a human).

"So I guess that's it." Tom shrugged. "We can't get one."

"Uh, you guess wrong," I said, glowering. I took a break for reconnaissance and spotted a little soccer boy headed toward the men's room, so I jumped out the door and leaned against the building, trying not to look suspicious. When the coast was clear, I slipped back inside. "Now get back up on that garbage can and get me a bird," I bossed like a wife in a bad sitcom. Tom sighed and dutifully climbed back up on the can as I held it steady it with all my strength—I kept picturing the can skidding out from under him and Tom dangling from the smelly bathroom ceiling, broken-armed and clutching a starling chick. We had repositioned the can so the lower roof ledge was smooshed right into Tom's armpit. "Hold out your hands," he told me, and into them he dropped the tiniest, ugliest, most unpromising little creature the earth has ever brought forth.

I'd raised dozens of chicks of many different species, from hummingbirds to red-tailed hawks, and of course the several starlings. But until now I'd never seen a baby bird that was actually *wheezing*. Like all songbird nestlings, this chick was mostly beak, with a big, fleshy orange gape designed to serve as a target for adult birds: *Drop food here.* When a chick is stimulated by movement and sound, the gaping response is induced. Wanting to make sure this bird possessed some tiny semblance of health, I tickled the bill and chirped like a starling; the little bundle threw back its head, and the bill popped open 180 degrees. Perfect.

This chick was only five or six days old and would require constant care: a steady temperature of 85 degrees until its feathers grew, and feedings every twenty minutes, dawn to dark. I had hoped to rescue a bird that was a few days older, one that was still young enough to tame but already raised into a bit more size and strength by its real bird parents; I wished I could put this one back to cook a little longer. But the nest was doomed, and with the arousal of my maternal instincts inspired by the gaping experiment, I was already starting to bond with this sad little chick—I couldn't bring myself to return it to the nest to be swept away with its ill-fated siblings. I knew I should get another chick to help keep this one the proper temperature and to increase my chances of ending up with one living, healthy starling for my research—baby birds, captive or wild, are unsettlingly ephemeral, subject to respiratory infections and weakened by ectoparasites of the sort I already saw crawling on this chick's bare skin. At this new request, Tom said—firmly—"No fucking way." He couldn't and *wouldn't* attempt to nab any more chicks. I opened my mouth, then wisely closed it again.

So that was it. This was our starling. I could feel the naked, translucent-skinned belly hot in my palm as the bird slept with its head drooped on my thumb. I tucked the chick carefully into my handy baby-bird incubator—my cleavage—and the three of us went home.

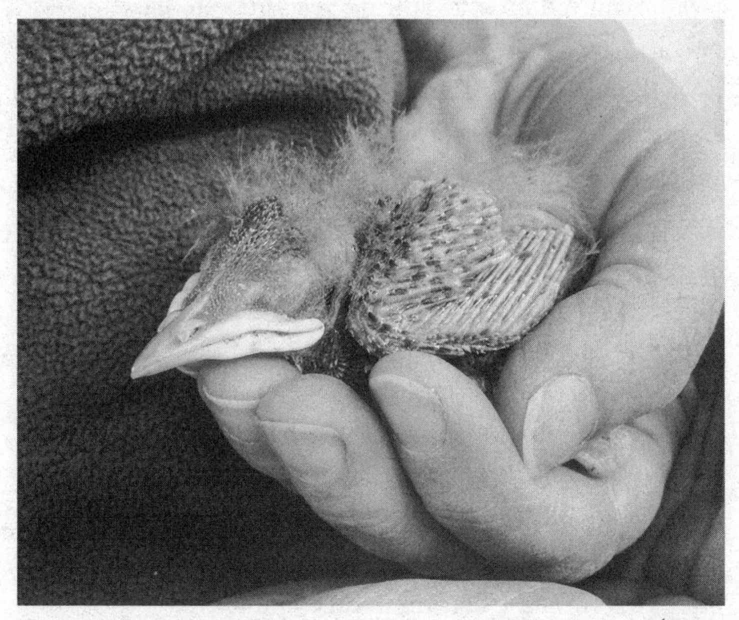

Our sleepy starling chick on the day we brought it home. *(Photograph by Tom Furtwangler)*

It was at this point that I morphed from "Lyanda the Innocent Citizen Removing a Nonnative Bird from a Public Space" to "Lyanda the Starling Outlaw." As it turns out, you may torture, maim, or murder a starling, but in Washington State, as in many states, you may not lovingly raise a starling as a pet. One of the ostensible reasons given by wildlife officials I spoke with was the prevention of propagation. There are already too many starlings, and people raising them as pets might eventually release the captive birds, making things worse. Something like this happened in the case of the house finch, a native bird with a geographic range that was once limited to the west side of the Rockies. The males have bright red breasts, sing all year, and are easy to keep, which made them marketable pets. In the 1940s, finches were illegally netted along the West Coast and transported east, where they were considered exotic and became popular. When there was an official crackdown on the wild-bird-pet trade, hundreds, perhaps thousands, of finches were released in New York by dealers seeking to avoid charges. The birds quickly acclimatized and eventually spread across the east side of the continent.

In the case of the starling, though, that rationale doesn't hold up. For one thing, the species has already overrun the country; it would take a huge number of released or escaped starlings to effect a noticeable increase in their population. On the contrary, it is far more likely that the removal of just one chick from the outside world could decrease the future starling population by scores, possibly even hundreds, of birds. (Starlings are able to reproduce at nine months old and often raise two broods a year. Say our bird fledged just three young its first breeding season, then those young, and all their future young, fledged three young each year... the numbers scale up quickly.) I'm not suggesting that starlings are a good pet choice for most people, but I do think the current standard makes little sense. In my opinion, if starlings remain legally unprotected, then we ought to be permitted to raise orphaned starlings in our living rooms.

It took just a few minutes to get our new chick from the park bathroom to its new home. I'd already prepared a mix of crushed dry cat food, hard-boiled egg, applesauce, calcium, and avian vitamins, with just the right balance of fat and protein for a baby starling. This I proffered in tiny bites at the end of a wooden stirring stick pilfered from Starbucks. (Baby bird, stirring sticks . . . my petty-theft rap sheet was growing by the hour.)

Though the bird was a decent eater, it remained sneezy and parasite-ridden. We hesitated to give it a name, not wanting to personalize our relationship and become more attached than necessary to what might be a transitory little life. Besides, we didn't know if it was a male or a female, so picking a name would be tricky.* Tom sometimes called the chick "little buddy," but overall we stuck with "it."

* It is difficult to sex starlings accurately before they reach breeding age, the first spring after hatching. Often the irises of female birds are more defined around the pupil than those of young male birds, but this measure is subjective and only about 70 percent accurate. I used calipers to measure Carmen's skull, which was in the female range, but ranges overlap—a large female skull can be larger than a small male skull. Once the birds acquire their breeding characteristics, things change. Males have longer,

For its first several weeks with us, the chick lived on the desk in my writing studio. Its nest was a plastic cottage-cheese tub lined with paper towels—I kept a roll handy so I could change them often. The tiny black ectoparasites that jumped off its thin skin were easy to spot against the white paper towels. I picked the nits up with tweezers and squished them. Keeping the makeshift nest clean wasn't difficult; most songbird-nestling poo is encased in shiny, strong fecal sacs that the parent birds remove with their bills and drop over the edge of the nest, so there is not much mess. I just plucked these poo sacs out with my fingers. And like other nestlings with an unconscious evolutionary imperative to keep a clean and disease-free bed, our chick, when it got a bit older, hung its rear over the edge of the nest and let its poo drop outside—theatrically heaving its tiny bum to the plastic rim, wiggling its still-featherless tail back and forth, and letting loose its impressive dropping with seeming satisfaction before falling into one of its deep baby-bird slumbers.

Eat, poop, sleep. It reminded me of having a newborn human baby, and in some respects it was even more restricting. The metabolic needs of an unfeathered chick are high and constant. Watching backyard nests, we can observe how frequently the parent birds come and go, bearing wriggly gifts of protein-rich insects and larvae for their young. As the stand-in parent bird, I had to feed my chick several times an

---

shaggier plumage on their shoulders and a punky look to their neck feathers, which are raised during singing and display. The bases of the bills also change color during breeding season and match our cultural stereotyping: girls' are pink, boys' are blue.

hour. When my daughter, Claire, now a teenager, was a baby, I could at least wrap her in a blanket and take her out with me—with this bird, I could barely leave the house. One day I decided to try packing its nest and food and bringing it along on my errands, planning to feed it as I went, but the chick quickly got too cold away from its heat lamp, so I ended up having to shop for groceries that day with a baby bird nestled, once again, between my breasts for warmth. For the most part, I was stuck at home. If I happened to be away overlong, the *Feed me!* baby-starling chirring sounds that poured forth from the tiny chick filled the entire house.

Always hungry. *(Photograph by Tom Furtwangler)*

At a couple weeks old, the chick started getting more mobile, and though it stayed on my desk, I put its nest inside a ten-gallon glass aquarium. It would jump out of the plastic tub and slump around the floor of the aquarium, its cartilaginous legs not yet able to hold it up, but it still preferred to sleep in its nest. This it accomplished extravagantly, with its head hanging over the rim and breathing with that hot, baby-bird heaviness (the cottage-cheese tub really was getting too small, but the chick seemed attached to it and chose it over the larger Tupperware nest I offered). When the little bird's legs began to ossify and get stronger, I added a low perch to one end of the tank, and it loved to jump on and off the stick and practice balancing there. But right from the start, its favorite place to play, sit, and sleep was on me. Tucked in my hands or on my lap, under my shirt.

Baby songbirds are not downy like baby pheasants or chickens or shorebirds or any of the other so-called precocial chicks that are born ready to run about. They are naked at hatching. Horny, sheath-covered pinfeathers emerge during the first week and take shape over the course of several weeks—the birds "feather out," as the ornithologists say. With its prickly pins, our chick felt a bit like a hedgehog. But soon the little starling's pinfeathers unfurled. It became as soft as a bunny and could stay warm more easily; now it liked to snuggle into the crook of my elbow and—especially—on my neck, under my hair.

Our tuxedo cat, Delilah, was only too happy to help oversee the care of the baby bird. She affected a great nonchalance, which fooled no one, and sat on my desk along with the chick, my laptop and me between them. Occasionally Delilah would lift her paw ever so slowly, and when I glared at her she'd pretend she was just about to lick her toes and do some face-cleaning. I never left Delilah alone with the chick, and because she is good at opening doors, I had to pound a nail into the molding by the doorknob so when I left the room, I could pull a thick rubber band around the knob and over the nail to keep her out. Once I forgot and when I came back to the study, there was Delilah sitting right over the chick, their faces just inches apart. Delilah was purring.

Meanwhile, my constant care and parasite-picking seemed to be paying off; the bird was flourishing. After four weeks, the shape of the iris provided the first indicator of the chick's sex, and we gladly replaced the neutral pronoun *it* with *she*. We named the starling Carmen, which in Latin means "song."

Two

MOZART AND THE MUSICAL THIEF

Raising a baby bird is harrowing. It's difficult to duplicate the perfect conditions of a nest, and at any moment, something can go wrong—a slight variance in temperature one way or the other can cause a naked nestling to freeze or die of heat exhaustion; the lack of an essential ingredient in the diet can cause a failure to thrive and seemingly sudden death; or a bird might just be sickly, as Carmen appeared to be, and not survive chickhood. The night after we stole-rescued our baby starling, I had a nightmare. In it, I walked up a dream-twisted staircase, through a doorway, and into my own house. The bleeding bodies of almost-dead starlings covered the floor. I woke up shivering and shook Tom. "Oh God, oh God, oh God. Tom. This was a horrible mistake." Tom rolled over without breaking his snore. I threw on a robe, ran barefoot to my study, and shone my iPhone flashlight on the chick. I watched her breathe heavily. I checked the thermometer—a perfect 85 degrees beneath the warm red-light lamp. I reached in and felt the chick's body, picked an errant nit. Then I pulled up a chair and watched the baby bird breathe until morning.

More bright-eyed by the day, but still vulnerable in the early weeks. *(Photograph by Tom Furtwangler)*

Hovering constantly over Carmen in her early weeks, I envied Mozart, who'd had a pet starling but had skipped the angst of raising a chick. The bird vendors of Vienna did not sell their birds until they were sturdy and grown, and because it appears that Mozart's starling was singing a solid song on the day he bought it, we know that the bird had to be a full adult, probably at least a year old. Younger birds practice songs and mimicry, but few are accomplished enough to sing a line from a Mozart concerto. And though it is impossible to be sure of the minutiae involved in the procuring of Mozart's starling, we do know many essentials, including the lively time line.

April 12, 1784, Innere Stadt, Vienna. Mozart sat at the small desk in his apartment, dipped his quill pen, and entered the lovely Piano Concerto No. 17 in G in his log of completed work. This was Mozart's 453rd finished composition; he was twenty-nine years old.

May 26. Mozart received confirmation from his father, Leopold, that the fair copy of the concerto he had sent by postal carriage had arrived safely in Salzburg. Wolfgang wrote back that he was eager to hear his father's opinion of this work and of the other pieces he had sent; he was in no rush to have them back "so long as no one else gets hold of them." Mozart was always a little paranoid that his music might fall into the wrong hands and be imitated or outright stolen by a lesser composer.*

As for what happened next, there are many possibilities. But it might have gone something like this:

May 27, Graben Street. Mozart's stockings pooled in wrinkles around his ankles, and he paused on the bustling roadside to pull them up. As he tucked the thin silk under his buttoned

* There are scores of Mozart biographies in the world, and though they typically agree on the known facts, they all provide varied, contradictory views of the composer's personality. When veering into matters having to do with Mozart's nature or inner life, I focus as much as possible on what I have been able to glean from his own words as they appear in his hundreds of published letters and relied primarily on two fine translations, which I use interchangeably in this book: *Mozart's Letters, Mozart's Life,* with letters selected, edited, and translated by Robert Spaethling; and *Wolfgang Amadeus Mozart: A Life in Letters,* with selected letters edited by Cliff Eisen and translated by Stewart Spencer. I recommend both volumes highly.

cuffs, he was startled by a whistled tune. It was a bright-sweet melody, a fragment beautiful and familiar. It took Mozart a wondering moment to recover from the shock of hearing the refrain, but when he did, he followed the song. The whistles repeated, leading him down the block and through a bird vendor's open shop door. There, just inside, Mozart was greeted by a caged starling who jumped to the edge of his perch, cocked his head, and stared intently into the maestro's eyes, chirping warmly. This bird was flirting! If there was one thing Johannes Chrysostomus Wolfgangus Theophilus Mozart responded to, it was flirting. Then the starling did it again; he turned away from the composer, pointed his bill skyward, fluffed his shimmering throat feathers, and sang the theme from the allegretto in Mozart's new concerto, completed just one month earlier and never yet performed in public. Well, he *almost* sang the tune. The starling made a minor rhythmic modification (a dramatic fermata at the top of the phrase) and raised the last two Gs in the fragment to G-sharps, but the basic motif was unmistakable.

The starling's mimicry is not surprising in the least—as birds in the mynah family, starlings are among the most capable animal mimics on earth, rivaling parrots in their ability to expertly imitate birds, musical instruments, and any other sounds and noises, including the human voice. But how did the starling in the shop learn Mozart's motif? The composition was meant to be an absolute secret, not slated for public performance until mid-June, when it would premiere under Mozart's direction with the gifted young student for whom it was written, Barbara Ployer, at the piano.

Mozart was so delighted by the starling he almost forgot his shock. He and the bird whistled phrases back and forth, sharing snippets of their repertoires. Then Mozart pulled out his pocket notebook and copied out the bird's species name, *Vogel Stahrl,* a version of the German name for the bird referred to as the European starling in North America and the common starling in England.* One commentator claims that Mozart named his bird Star, a misreading of his note that simply referred to the species. Even so, it is handy to employ a moniker in telling a story, and as there is no record of the bird's actual name, Star will do nicely.

This story is not well known in its details, and some musicologists, acquainted with only the surface of the tale, claim that Mozart must have responded in a jealous fury to the bird's pirated rendition of his own composition. But when we look into the composer's pocket notebook, we see that nothing could be further from the truth. Beneath the words *Vogel Stahrl,* Mozart wrote his own version of the tune, then the starling's version.

Mozart's motif.

The starling's song.

* Later, Mozart would refer to the bird by the more common spelling *Vogel Staar.* Today in German the species is typically referred to as *Vogel Star.*

His comment on the starling's interpretation? *Das war schön!* "That was wonderful!"

It would not have been at all odd for Mozart to keep a bird. Pet birds were popular in eighteenth-century Europe, part of the natural-history trend that characterized Enlightenment attitudes in polite society. Facilitated by an emerging international shipping trade, exotic birds such as parrots and mynahs, as well as animals ranging from wombats and kangaroos to great land tortoises, made their way into public menageries and into increasingly popular animal- and bird-merchant shops. ("Can peace be gained until I clasp my wombat?" Dante Gabriel Rossetti wrote to his brother in 1869 while waiting impatiently for his new pet to make its way across the sea from Australia to England.)

Exotic birds were expensive. In *The Georgian Menagerie*, cultural historian Christopher Plumb writes that a parrot could cost as much as a typical servant earned in a whole year, and bird-selling was good business for high-end shopkeepers who could afford to have the exotic species shipped in from Africa and Australia. But it was the trade in native birds such as chaffinches, bullfinches, doves, and sometimes starlings that made pet birds accessible to a wider population, bringing both decorative and musical interest to the middle-class salon.

Little is known about the local bird catchers, many of whom lived in near poverty at the fringes of society. They would catch, raise, and sell birds to vendors with proper shops, or sometimes they would sell the birds themselves, along with simple homemade cages, from seasonal street stalls rented with their last pfennigs. These were often family ventures, with tatterdemalion youngsters sent into the fields and woodlands to check progress on nests and eggs. Nestlings were pilfered and raised until they were grown, sturdy, and ready for sale. Though the work was socially unrespected, it was not unskilled. Local bird catchers might have been functionally illiterate, but they had to be accomplished natural historians, knowing how to identify and name species, find nests, and monitor the laying of eggs and the fledging of young. They had to know how to hand-raise birds, diagnose health issues, and sometimes cure them. They had to be thieves, scientists, veterinarians, and businesspeople, all at once. And yet, as Plumb points out, most of what we know of these tradespeople comes from court records in which they are accused of drunkenness, robbery, or petty crimes. It seems they were never considered part of the society in which the birds they raised found homes.

It was surely one of these skilled ruffians who hand-raised the starling Mozart chose before it arrived at the shop; the bird was tame and friendly, and the practiced shopkeeper had no trouble catching it and depositing it in a small wooden box lined with natural grasses that Mozart carried home to his wife, Constanze, whistling all the while.

Mozart's walk was a short one, but the noonday streets were bustling with horses, wagons, and hackney carriages. Several of the city's many homeless dogs brushed his legs, but they ignored the maestro and his mysterious box, intent on getting to the stalls of the street vendors who migrated from the suburbs each morning with their offerings of eggs, meats, cheeses, and wines; a well-behaved dog who sat quietly would get plenty of scraps. There was high-piled hair and the flouncing of hoopskirts, now in their last decade of popularity. There was the fragrance of roasting chestnuts and the smoke of kitchen fires and the manure from the carriage horses. There was, occasionally, the song of a street musician. On a normal day, Mozart had an eye and ear for all of this life—life everywhere was a thing to be drunk up and poured out again in his music. But this day he took no notice of anything. His mind was all on the little box. Mozart whispered to the bird within, maybe telling him about his new home. Meanwhile, Star, who had loved this man's voice in the shop, was now huddled into the darkest corner of his tiny crate, wide-eyed and silent. He was tame, yes, but no starling likes to be stuffed into a coffer and carried about. The poor bird was terrified.

Soon, Mozart reached his apartment at 29 Graben, a fashionable address—the Graben was then, as now, the central shopping and fashion district. The Mozarts' rooms were not spacious, but it was just the two of them, Wolfgang and Constanze, along with their small dog, Gauckerl, and now Star. Perhaps Wolfgang thought this new bird might bring a cheering presence to the house. The couple's first baby, little Raimund Leopold, had died the previous year when he was just six weeks old. He had been in the care of a wet nurse while Constanze and Wolfgang were in Salzburg visiting the elder Leopold—Mozart's attempt to sow goodwill between father and wife (though Leopold had never met Constanze, he was against the match from the start). The couple had left Raimund fat and happy, and Mozart blamed the child's death on their decision to raise him on breast milk rather than water and coarse-milled oats, as was commonly (and disastrously) recommended by the medical men of the era. That May afternoon when Wolfgang turned up at the Graben apartment with his starling in a box, Constanze was five months pregnant with their second baby. The child would be named Carl Thomas, and of the Mozarts' six children, he was one of just two who would survive to adulthood (it sounds shocking and sad, but this survival rate was a bit above average).

In my imagining, Constanze was bemused and also a bit put out at the new housemate (what pregnant woman needs something else to take care of?), but she could not have been surprised. She knew that her husband had been fond of pet birds from childhood—singing canaries, mostly. Any consternation she felt was dispelled by Wolfgang's unabashed joy. Star was unnerved by his short journey but settled quickly into his new cage, as these intelligent birds do, without much thrashing about. Wild foods and seeds were typically sold at bird shops, but most likely Star simply shared the family diet, feasting on bits and leftovers from the Mozarts' table, fed to him by hand or left in his cage. Starlings are omnivores, and the varied scraps from an eighteenth-century middle-class Viennese kitchen—meat, potatoes, fruits, and plenty of pastries—probably offered just the right fat-protein balance for little Star. (Carmen loves to nibble our leftovers; her favorites include lentils, spaghetti, and couscous salad.)

It is not known whether Constanze had any childhood pets, but since her father, Fridolin Weber, was a restive jack-of-all-musical-trades, their lifestyle was probably too unsettled for animals. Constanze, who grew up in the cultural and intellectual center of Mannheim, was the second of four Weber sisters, all of whom had classically trained voices. During Constanze's teen years, the family moved frequently to promote the singing career of the eldest sister, Aloysia.

Mozart was born and reared in provincial Salzburg but traveled widely throughout Europe as a child prodigy, performing on the violin and pianoforte alongside his sister Maria Anna (always called Nannerl), a brilliant pianist in her own right. The children were usually accompanied by both parents on these long and expensive journeys, which were fraught with the many dangers of carriage travel: poor roads, inclement weather, exposure to disease. Wolfgang was often sick and near death more than once. His short stature was the subject of public and medical comment and a concern to Leopold. Ill health would plague him always.

There came a time when Leopold could no longer shirk his duties as *Kapellmeister* to the prince-archbishop of Salzburg in order to parade his young prodigies about Europe. So, beginning in 1777, Mozart's mother, Anna Maria, chaperoned the twenty-one-year-old Wolfgang on a sixteen-month journey without Leopold (Nannerl stayed behind with her father to look after the household). The first months were spent in Mannheim, and then, at Leopold's urging, mother and son continued to Paris. While Wolfgang gallivanted about the city, teaching, composing, giving recitals, dandying, and attempting to ingratiate himself with potential patrons among the royalty and aristocracy, Anna Maria, who could not go about unaccompanied in polite society, languished in their dank rooms. "The stairs are so narrow," she wrote to Leopold, "that it would be impossible to carry up a klavier. Wolfgang cannot compose at home. I never see him all the long day and shall forget altogether how to talk." She fell ill in Paris and died fairly quickly, a tragedy from which Mozart's spirit never fully healed.

Young Mozart, alone in Paris with the body of his mother, could not bring himself to tell his father and sister what had happened. He lied in a letter to Leopold: "I have to bring you some very distressing and Sad news, which is the reason why I couldn't reply earlier to your most recent letter.... My dear Mother is very ill—she was bled, just as she had it done always, and it was indeed necessary; afterward she felt somewhat better—" He engaged a friend in Salzburg to break the news to his father and didn't write to Leopold with his own version of the truth for over a week. "I hope that you and my dear sister will forgive me for this small but necessary deception—when I thought about my own pain and sadness in relation to how it might affect you, I simply could not bring myself to overwhelm you with this distressing news." Vestiges of guilt and worry later emerged in his doting, anxious concern for his wife (whom he could not bear to leave), for his children, for his dog, and, yes, for his starling.

There is a heartbreaking oil portrait of the Mozarts that was commissioned after Anna Maria's death. The two grown children sit at the fortepiano; their father, Leopold, stands in shadows with his violin; and their beloved mother appears behind them in an oval-framed painting, her hair piled high and wide and wound with a blue ribbon. This portrait of the family is a powerful and ghostly presence at the Mozart Geburtshaus now, hung in the back of the dark, windowless, wood-rich room where Wolfgang was born. It is a bit discomfiting to explore the rest of the exhibit with the family watching—rustling and whispering and mourning there in the corner.

Mozart family portrait. *(Johann Nepomuk della Croce, 1781)*

After his mother's death, Mozart continued to travel extensively, but Salzburg was home base. As Mozart's genius crystallized in his early twenties, the town began to feel too provincial to hold him. With difficulty, he tore himself away from his fretting but highly intelligent father.

Leopold is misunderstood in the modern Mozart mythology. His difficult points—officiousness, anxiety, codependency, and condescending micromanagement of his family's activities—are well documented; his controlling nature is so over the top that it's almost funny, and there are thousands of examples of it in the family correspondence. When Anna Maria and Wolfgang were off traveling, Leopold wrote constantly, instructing his fully competent wife on the minutiae of business and life.

> *Wherever you are, always make sure that the innkeeper puts the boot-trees in your boots.... The music can always remain in the front in the trunk, but you should buy a large oilcloth and use both this and the old one to wrap it well, in order to ensure that it's really safe.... I shall send fresh socks by postal carriage.*

But Leopold loved his family rashly and dearly. He thoughtfully homeschooled his children, not just in music but in all subjects. He was a fine composer and a well-known violin pedagogue throughout Europe. Without Leopold, we would never have heard of Wolfgang Mozart.

Even so, the relationship between father and son would

always be fraught. As Mozart grew into a young adult, Leopold could not keep from insinuating himself into every aspect of Mozart's life. No matter where Wolfgang traveled, Leopold would send letter after disapproving letter, insisting that Wolfgang find more ways to ingratiate himself with the aristocracy, improve his connections with famous composers, and, always, make more money; the letters were full of detailed advice on exactly how Wolfgang should go about accomplishing all these things. His love for Wolfgang shone through without fail, yet he could not help constantly reminding his son how much the family had laid out for travel and clothes and lodging in service to Wolfgang's genius. He deployed a complicated cloud of guilt, love, and indebtedness that followed his son everywhere, and always.

After his wife's passing, Leopold became even more clingy, anxious, and controlling, and Wolfgang's desire to leave Salzburg did not help. Mozart provoked his own dismissal from his underpaid employ with Salzburg's Archbishop Colloredo and fairly ran away to Vienna, leaving his talented sister behind in the throes of depression. Nannerl was now the keeper of her father's household and knew she had only two options open to her: live as a respectable spinster, or marry. Both options required that she give up her life as a musician. Leopold was focused entirely on Wolfgang and no longer promoted his daughter's talents. She spent days in bed, suffering under the stark truth of what her life was to become. (Eventually she would marry, though not happily.) The spectacularly talented Nannerl stopped playing the pianoforte.

When Wolfgang and his mother were in Mannheim, they met the musical Weber family. Mozart scarcely noticed Constanze, besotted as he was by the eldest sister, Aloysia, with her fashionable beauty and her diva's soprano. Mozart concocted a wild plan in which he would run away with Aloysia to Paris and compose arias for her pure voice that would make them both famous. He wrote all about it to his father: Would Leopold tell him what a *prima donna* earned in Verona?

Poor Leopold! When he read Wolfgang's long missive outlining the naive scheme, he fell into absolute fits. "My Dear Son," he wrote, "I've read through your letter of the 4th with bewilderment and shock." Leopold claimed to be so distressed that he had not slept the entire previous night and as a result was so exhausted that he could hardly write and struggled with each word. This did not stop him from composing a letter that was dozens of pages long, expounding in great detail on the folly of his son's plan, which seemed, to Leopold, like a far-fetched fantasy that would make social lepers of the entire family. "How could you allow yourself even for a moment to be taken in by such an appalling idea... to cast aside your reputation—your old parents, your dear sister?... To expose me to mockery and yourself to contempt?" Finally, he turned to his favored method of twisting the knife: guilt. "Remember me as you saw me when you left us, *standing wretchedly beside your carriage;* remember, too, that, although a sick man, I'd been up till 2 o'clock, doing your packing, and was at your carriage again at 6, seeing to everything for you—afflict me now if you can be so cruel!"

But in the end Leopold needn't have worried, at least not about the eldest Weber daughter. Aloysia swiftly jilted Wolfgang and married the more mature, financially solvent (and much taller) Joseph Lange, an actor, singer, and portrait painter. Mozart, meanwhile, traveled all over Europe, composing and performing, and eventually returned to Vienna, where the Webers now lived. Herr Weber had since passed, and Frau Weber was taking in boarders to help make ends meet. Mozart roomed at the Weber home for some weeks, and during this time he tidily transferred his affections from Aloysia to her sister Constanze. Wolfgang's affections for Constanze might have been less youthfully wild than his infatuation with Aloysia, but they were sincere. He intended to marry her.

Leopold was in a disapproving tizzy over the impending nuptials. For years he had been meticulously plotting the course of his son's fame. Now Wolfgang wanted to derail his own chances for renown and esteem by marrying? And into a family whose name meant nothing, who had no money, no prospects, no sons to ensure future income? Leopold despised the lot of them, sight unseen. But Wolfgang was twenty-five years old and ready to settle down. He was comfortable with the Webers, and through simple proximity, he and Constanze had developed a dear friendship and then, over the months, an intense affinity. He wrote to his father with trepidation but was firm in his resolve:

> *The middle one, my good, dear Constanze, is the martyr of the family, and probably for that very reason, is the kindest-hearted, the cleverest, in short the best of them all.*

Hoping to appeal to Leopold's concerns over economy, he emphasized Constanze's practical virtues:

> *I must make you better acquainted with the character of my beloved Constanze;—she is not ugly, but also not really beautiful;—her whole beauty consists of two little black eyes and a graceful figure. She has no great wit but enough common sense... she is not extravagant in her appearance, rumors to that effect are totally false;—to the contrary, she is in the habit of dressing very simply... and most of the things a woman needs, she can make herself; indeed, she does her own hair every day.* She knows all about householding and has the kindest heart in the world—I love her and she loves me with all her heart—now tell me whether I could wish for a better wife?*

Constanze had plenty of wit. She possessed an artistic spirit and a solid temperament. And in spite of the largeness of her husband's personality, she held on to a sense of bright independence. She traveled and managed parts of the household music business. Mozart wrote songs for her lovely soprano voice. She governed the couple's ever-changing financial situation as well as anyone could and maintained relative equanimity amid the chaos of composing, parties,

* Doing one's own hair might not sound so impressive, but it was common even among the middle classes to have a friseur make home visits to do up the high styles of both men's and women's hair or wigs.

recitals, pregnancies, children, and the labors of middle-class eighteenth-century domestic life in dusty Vienna. Though Leopold was predisposed to find fault, even he commented on Constanze's commonsense home economics after his visit to the young couple's apartments. Wolfgang and Constanze's marriage was not without troubles, but it was a sweet one, and happy overall.

Star joined the family in the middle of the marriage, during the most musically productive, prosperous, and engaging years of Mozart's life. He might have been the smallest member of the household and is barely mentioned in most biographies, if he makes it in at all, but the starling is never far from the center of Mozart's unfolding story. Any Mozart historian would give an arm for this bird's-eye view of these years. Star's vocal acrobatics accompanied the composition of at least eight piano concertos, three symphonies, and *The Marriage of Figaro.* He was present for Leopold's ten-week visit to the young couple's house, the only visit the elder Mozart would ever make. Star heard, and likely joined in singing with, the debut of the Haydn Quartets, performed in the parlor with Papa Haydn himself in attendance. Star was present in the house during the birth of Carl Thomas, in 1784, and Johann Thomas Leopold, in 1786. He witnessed, with his inquisitive starling's eye, the mourning in the household when tiny Johann Thomas died at just three weeks old. Star has been considered a footnote to the Mozart biography, but after living with a starling, I have become convinced that the bird brought a constant current of liveliness, hope, and good cheer into these complex years, one that sustained Mozart's heart and music.

Three years after Mozart brought Star home, his father, Leopold, passed away, leaving his son with a knotted mixture of guilt, mourning, and relief. Mozart did not travel to the memorial in Salzburg, where Leopold was buried without mourners. Mozart's starling died just two months later, and in honor of the bird, Mozart organized a formal funeral, donned his most elegant finery, recruited friends as velvet-caped mourners, and penned an affectionate elegy. My favorite translation is Marcia Davenport's, from her 1932 biography of Mozart, now out of print; it captures the simultaneous jocularity and formality of the little verse. After a few lines that announce the starling's death, Wolfgang laments:

Thinking of this, my heart
Is riven apart.
Oh reader! Shed a tear,
You also, here.
He was not naughty, quite,
But gay and bright,
And under all his brag
A foolish wag.

The poem shows that Mozart had become thoroughly acquainted with the typical starling personality—bright, personable, charming, mischievous. Some historians have claimed that the funeral verses are simply a farce, but no one who has lived with a starling would dream of making such a suggestion.

Three

UNINVITED GUEST, UNEXPECTED WONDER

When Carmen was four weeks old and flapping about her aquarium, I realized that she needed a larger home. We moved her downstairs into a big cage on wheels. She still spent hours each day outside the cage, cuddling with all of us and practicing her emerging flight skills. Starlings are insatiably social and Carmen would call piteously when we were out of her sight. When she needed to be in the cage (on hot days when we had to have the windows open, or when we were cooking in the kitchen with dangerously boiling water, or when we were eating and wanted to keep her out of our food), we just wheeled it to wherever we were in the house so she could see us and chatter with us.

By about this age, starlings have pretty much finished growing. They get, if anything, a little smaller as they mature, shedding baby fat and acquiring an adult sleekness. But they do get busier and more active. Though Carmen's cage was the largest I could find, I couldn't imagine keeping her there for long. My super-handy dad, Jerry, came to stay for a few days to help me design and build a custom aviary in the corner of our mudroom—a room within a room. I made sure Carmen's cage was in view as we worked so she could observe our progress on her new home. Maybe if she watched it being built, it wouldn't seem too big and ominous when she moved in.

We made a sturdy frame with two-by-two-inch raw-cedar posts, used lighter one-by-twos for struts, and neatly stapled metal-grid "hardware cloth," as it's called, to the outside of the structure (rather than the inside, to keep Carmen from getting her toenails stuck in the staples). We fit the whole enclosure in the corner of the mudroom, a spacious throughway between the kitchen and dining room, so that Carmen could see us and hear us in the places where we spent most of our time. The aviary reached from the floor to the ceiling, and the back wall ran along a huge window with a view of trees, birds, and sky. The raw wood and metal had a pleasing, natural look, and I added tree limbs for perches.*

It took two days of dawn-to-dark cutting and pounding, several trips to the hardware store, and lots of swearing on the part of my dad, but finally the beautiful aviary was ready. Jerry and I surveyed our handiwork, high-fived, and poured a celebratory beer. But Carmen was wary. Starlings are like cats—though they are brave and inquisitive, they also like routine, and this was a whole new world.

---

* It's important to do some research before using natural limbs in a bird's home, as some bark is toxic to certain species.

Carmen full grown at six weeks, and feathered out in her gray juvenal plumage. *(Photograph by Tom Furtwangler)*

While Jerry finished his beer, I decided it was time to introduce our little bird to her new home. I stood in the aviary's doorway with Carmen on my shoulder. Her eyes grew large. I stepped inside and stayed there with her among the branches so she could get her bearings. I could feel her feet tighten uncertainly on my shoulder, but after about fifteen minutes I gently coaxed her onto a branch and slipped out the door. She squeezed into the highest corner and stayed there, silent and overwhelmed, for about an hour. The next hour she began to explore quietly. By the third hour Carmen had taken full possession of her aviary and was hopping confidently from perch to perch.

There are tiny mirrors and all kinds of toys that I switch out daily to keep her interested (her favorites are empty milk cartons on the floor that she can overturn and stomp on and plastic bottles that she can practice rolling around on). There is lots of room to fly and explore. As frequently as possible, I leave the door open so she can come and go as she likes, but even then she will often just hang out in her aviary, sometimes flying between my shoulder and her favorite branch and back again. "Good flying!" I tell her, wanting to encourage healthy exercise. The aviary is her home, her safe place. When she is alarmed, or bored, or sleepy, or preening after a bath, that is where she wants to be. And when she steals things from the household that she is not supposed to (like thumbtacks or money), that is where she goes to hide them.

We designed the door to be big enough for humans to walk through, for ease of cleaning, and as we worked, we decided to make it a Dutch door, two half-doors, one on top of the other. We thought this would make the door less unwieldy and help it maintain its integrity on the hinges over time, but it turns out to be one of Carmen's favorite things about her home—she likes to have the top door open so she can perch on the closed bottom door and survey her domain. She is a bit shy of new people, and this is where she prefers to sit—on her split-level door—when guests visit. Here she can get a good look at the interloper but still be able to make a quick getaway to the comfort of her favorite branch inside the aviary if she gets scared. Meanwhile, it is a nice place for guests to view her without looking through cage wires.

Soon after we finished the aviary, Carmen began her first molt, from her plain gray juvenal plumage to her first adult plumage. Every day she became more spotted and glistening.

One of the most frequent comments I hear about Carmen is "But she's so *pretty.* This isn't what all starlings look like." It seems that because starlings are so despised, they are also expected to be ugly, or at least plain. Carmen is more beautiful to me than other starlings are because I know her personality and have grown fond of her. But as starlings go, she is no great looker. If anything, she is a bit spindly, as her pectoral muscles are not well formed due to a lack of extended flight. She is tidy and clean, but so are all healthy young birds.

First fresh, starry adult plumage. *(Photograph by Tom Furtwangler)*

I think the thing that most surprises people about Carmen is her glitteriness. Like all starlings', Carmen's plumage is iridescent—muted black when seen from one angle but coming to shimmery life when seen from a slightly different one or in a certain slant of light. Starlings are painted like oil slicks, layered with shining purple, blue, magenta, and green. Iridescence in feathers is created through structural changes in the feather surface that make them appear vibrant at certain angles—microscopic bumps and ridges on the barbs and barbules refract and scatter light. The gorget of a hummingbird—garnet at one glance, brown at another—is the crown-jewel example.

When starlings molt in the fall, many of their fresh iridescent feathers are tipped with white, giving the birds their celestial pattern. But the structural changes that make starling feathers iridescent also give the feathers added strength, protecting them from extremes of light and weather. Without the reinforcing benefits of coloration, the starling's white tips wear off in the winter, leaving the birds all glistening black in the spring. It's a unique strategy for acquiring breeding plumage—most songbirds molt into bright new breeding feathers to attract a mate in the spring, but starlings simply wear away their white to come up with a glimmery new look for the season.* To birds, most of which can see on the ultraviolet spectrum invisible to humans, iridescent starling

* I've heard some naturalists question this, but Carmen was a perfect experiment. Her first complete autumn molt filled the house with clouds of feathers. (It is astonishing how many feathers are layered onto one little bird—about three thousand on a starling.) It was a nerve-racking few weeks for me. Every time I came home, Delilah would greet me at the door with feathers in her mouth, and I'd rush to Carmen's cage while searching the floor for blood and bird entrails, but she was always safe and happy. Like most cats, Delilah just enjoyed playing with feathers. Once I survived the psychological torment of the molt, I admired Carmen's fresh, starry breast. Come spring, all her wild colleagues had lost their white tips, but Carmen's, protected from the elements, were as pristine as ever.

Even in black-and-white, the details of a starling's plumage are wondrous. From top to bottom, this close-up shows the small covert feathers over the wing, the shorter secondary wing feathers, the long-lined edges of the primary wing feathers, and the pointed breast feathers. *(Photograph by Tom Furtwangler)*

bodies literally glow. Even if you don't have UV vision, a starling in sunlight is absolutely stunning.

It is intriguing that a bird so common, likely the bird most often seen by city dwellers, is so little understood, or even recognized. In spite of the starling's unique, sparkling plumage, the majority of people, including urbanites who live alongside starlings every day, cannot accurately identify them. Many confuse starlings with the much less sparkly blackbirds, or even with baby crows. (All three of these birds are in the large Passeriformes order of so-called songbirds, but other than that, they are not closely related, and baby crows, once they are walking about, are no smaller than adult crows.)

The starling species common in North America is the European starling, one of the Old World Sturnidae family, a group that includes more than a hundred species of starlings, mynas, and oxpeckers spread across Europe, Asia, and Africa. All starling species are shimmery, precocious, terrestrial, gregarious, vocal birds, all of them iridescent, many of them extremely colorful—shades of cobalt, magenta, and brightest yellow. The superb starling of East Africa is among the most stunning small birds in the world, with brilliant turquoise and black plumage and shining golden eyes. One of them stole a piece of my sandwich as I was picnicking beneath a tree overlooking Lake Manyara in Tanzania. At first, the starling seemed quite polite about it, stepping up and tilting his head to one side, looking at me with his sun-yellow eye. I half expected him to say, "May I please have just a bit of crust? Perhaps a nibble of that lettuce?" But then he jumped in and took a bite so fast I hardly saw it happen. The native range of the European starling extends across all of temperate Europe and West Asia, but the species has been introduced in Australia, New Zealand, South Africa, Argentina, and North America, and the birds have spread just about everywhere else except, thankfully, the neotropics, where their impact on delicate native songbird populations could prove devastating.

Regarding the presence of starlings in North America, some blame Shakespeare. In the 1800s, "acclimatization societies" began to form across the country, following successful models in France. It was a vulnerable time for many newcomers to America, who were homesick and hungry for the arts, literature, flowers, and birds of their homeland. The aim of the societies was to introduce European species that would be "interesting and useful" to the seemingly deprived New World species that would offer aesthetic and sentimental inspiration through beauty and song.

Eugene Schieffelin was a pharmacist who lived in the Bronx. He was an eccentric, an Anglophile, and a Shakespeare aficionado. Some say he was also an ecological criminal and a lunatic, but I would argue for a gentler description; perhaps "flawed." As deputy of the American Acclimatization Society of New York, Schieffelin, it is believed, latched onto the personal goal of bringing every bird mentioned in the works of Shakespeare to Central Park. Armed with his treasured copy of the exquisite *Ornithology of Shakespeare*, an 1871 volume in which James Edmund Harting assembled every allusion to birdlife in the whole of the Shakespeare canon, Schieffelin zeroed in on the Bard's single reference to a starling, in *Henry IV*. It is a decisive scene: King Henry commands that the willful soldier Hotspur free his prisoners, but Hotspur replies that he will do nothing of the kind until the king agrees to pay the ransom that will free Hotspur's brother-in-law Mortimer from the enemy. The king flies into a fury and forbids him to mention Mortimer's name. After the king's exit, Hotspur imagines a fanciful retribution, and here enters our star:

> *He said he would not ransom Mortimer;*
> *Forbad my tongue to speak of Mortimer;*
> *But I will find him when he lies asleep,*
> *And in his ear I'll holloa, "Mortimer!"*
> *Nay.*
> *I'll have a starling shall be taught to speak*
> *Nothing but "Mortimer," and give it him*
> *To keep his anger still in motion.*

Shakespeare was attentive to birdlife; larks, nightingales, and chaffinches wing and sing their way through the plays and sonnets, and in his unique *Ornithology,* Harting cataloged every one and quoted the lines in which they appear. The acclimatization societies did in fact try to bring in many of these species, but with the obsessive powers of a true eccentric, Schieffelin fixated on this one slender reference to the starling. Introductions of Shakespearean chaffinches, nightingales, and skylarks, and earlier efforts by Schieffelin to establish starlings, had resulted in nothing but cold, starved dead birds. Eugene resolved that his next starling attempt would not fail.

In 1890, he paid out a princely sum from his private stores (surely enough to satisfy Mortimer's ransom) to purchase eighty starlings from an English source, and he perhaps laid out a bit extra to ensure that they would be well tended on their long journey to the New York port, where Schieffelin met them in person, enlisting help from his houseman to carry their crates. He released his bewildered birds on a snowy March day in the middle of Central Park. I think of him there—gloved, worried, flush with hope and an honest, if misguided, love. The release could not have been all he'd envisioned. The birds would have been tentative in the cold and the snow, perching uncertainly in the leafless maples. This was not the romantic bursting into flight that Schieffelin had surely imagined. But eventually the birds lifted into the gray winter skies. Genetic research in sample populations across the continent leads ornithologists to believe that all of the two hundred million–some starlings in North America, including my little Carmen, are descendants of Schieffelin's birds.* (It's interesting to note that our starlings have quantifiably less genetic variation than starlings in their native European range. This is in line with what evolutionary biologists call the "founder effect," in which the number of animals introduced—in this case, Schieffelin's eighty-odd birds—is not large enough to contain all the genetic variation of the original population.)

The spread of the starling was swift and complete. The Central Park birds dispersed into an emerging starling Shangri-la.

---

* Though starlings are just as reviled in New York City as they are everywhere else in the U.S., this genesis story is honored in the Bronx with a Starling Avenue, a Schieffelin Avenue, and a Schieffelin Street, and it is rumored that one of New York City starlings' favorite roosts is Central Park's Shakespeare Garden.

They were accustomed to human presence and habitation in their home in England; the young city of New York would not have fazed them in the least. Some birds stayed close to Central Park; others flew to growing neighborhoods that provided warmth (sheltered buildings and perches above heated chimneys), food (human leftovers and leafy parks inhabited by tasty grubs and insects), nesting places (cavities created by building cornices and exhaust tubes), and ample foraging (grassy expanses in parks and gardens). Their progeny spread to other developing towns, first nearby, then farther and farther across the land. More descendants flocked to agricultural areas, where they easily found sustenance in the form of grain and fruit crops.

Starlings exhibit every characteristic of a successful animal invader: they are robust, aggressive, omnivorous, and unfussy about nest spots, and they reach sexual maturity at just nine months. They reproduce prolifically, with two clutches per season, sometimes more, and raise large broods of four to six chicks. (One clutch is the norm for most migrant songbirds, though nonmigratory resident birds in temperate climates—like robins and chickadees—will often raise two broods.) Starlings are inquisitive and intelligent, which makes them adaptable and ready to explore and colonize new places.

We know that curiosity killed the cat, and to make up for this, cats are granted nine lives. It is difficult to imagine a more brazenly curious creature than the starling, and to balance things out, it seems they ought to have nine thousand lives. People who live with starlings know this. There is a website administered from New York City called Starling Talk, where people who have starlings as pets gather to discuss matters such as the raising of baby starlings, starling health and diet, and the general wild craziness of life with a starling in the house. Nearly all the Starling Talkers came to have a starling because they found an injured or orphaned bird and, since starlings are unwanted by rehab centers, decided to take the care of the bird into their own hands. The discussion at the website is lively and, as with any obscure social coterie, often veers into the arcane and nerdy—matters that only other starling-keepers would care about or understand.

Starling Talk members have learned from sad experience that if you have a starling loose in the house, you must avoid leaving glasses of any liquid on the counter so that your bird will not lean in to drink it, get its wings pinned, and drown; close the toilet lid before flushing so the curious bird does not follow the entrancing swirl down the pipe; take care in using the garbage disposal; make sure before you turn on the microwave that your bird has not somehow slipped in; refrain from chopping vegetables with large knives, as starlings cannot help investigating with tiny bills and toes that are all too easy to inadvertently slice off. You have to watch your step—starlings are so completely at home with their human flock-families that they are constantly underfoot, and it's easy to accidentally trample their tiny, hollow-boned little bodies. One day I couldn't find Carmen anywhere, and finally, after about an hour of searching, I took a break, deciding she was probably napping somewhere and would turn up when she was ready. Besides, I was getting hungry. When I opened the refrigerator to get the peanut butter for my sandwich, she jumped off the shelf next to the eggs and rushed to my shoulder. She tried to shake the cold off her feathers, and I tucked her under my shirt to thaw. Poor little thing. I can't imagine how she jumped in there without my knowing, but other than being a bit chilled, she seemed fine.

I almost sweep Carmen into the dustpan pretty much every day. She likes to play right under the path of the broom. *(Photograph by Tom Furtwangler)*

Starlings bring this bright curiosity to the exploration of their world, and the only habitats in North America that starlings avoid are large expanses of wooded or forested areas, arid chaparral, and desert. Ornithologist Paul Cabe proclaims that given the starling's omnivorous diet and ability to make use of buildings as nest sites, no native bird in North America, not even the crow, is better adapted to the urban wilds than this invader. It took starlings just eighty years after their release in Central Park to populate the entire continent. Eugene Schieffelin lived to realize that his starling introduction succeeded, but he could not have predicted how the story would unfold over the next century. What would Schieffelin make of his triumph now? In *Tinkering with Eden*, Kim Todd suggests that he ought to have read his revered Shakespeare more closely. In *Henry IV*, "the starling was not a gift to inspire romance or lyric poetry. It was a bird to prod anger, to pick at a scab, to serve as a reminder of trouble. It was a curse." Perhaps even Schieffelin would realize that no matter how pretty the starlings were, how mesmerizing their vast autumn cloud-flocks, here was an experiment that had gone terribly wrong.

Since that fateful introduction, starlings have made poor guests as they spread across the country, and I regret to admit that Carmen is no exception. The first time we had to go out of town and leave Carmen for several days, I wasn't sure what to do. I could have just had someone come in to feed her and change the newspapers that line her cage, but since starlings are so social, I worried about her becoming neurotically lonely and pulling out her feathers, as isolated parrots do. My friend Trileigh Tucker is a geology professor, birdwatcher, naturalist, and Crazy Cat Lady (just five at the moment). I knew she would be the perfect starling-sitter, and Trileigh kindly (or maybe naively) agreed. I coaxed Carmen into Delilah's kitty-travel box and loaded her old rolling cage into the back of the Subaru—this would be her guest room. At Trileigh's house I gently held the kitty box and (remembering Mozart's journey with boxed-up Star from the bird shop to his home) whispered encouragement to the horrified Carmen, while

Trileigh and her tolerant partner, Rob, gingerly maneuvered the cage up their narrow stairway and into Trileigh's home office. Carmen settled in quickly and I left feeling reassured.

Upon our return, Trileigh informed me that Carmen's first act as a houseguest when she was let out of her cage was to fly to the aquarium, perch on its edge, and, quick as a great blue heron, pluck out a shining purple guppy. It happened "instantly," Trileigh marveled, before she had time to blink. Carmen always knows when she has taken something she isn't supposed to have, and she flew with her prize to the top of a bookshelf, out of reach. Trileigh yelled and waved her arms helplessly as Carmen swallowed the guppy. In spite of this bad behavior, Carmen was invited back, and Trileigh has claimed the dubious distinction of starling auntie.

Carmen was acting true to her species' form. Starlings have behaved atrociously in their New World. They feast in great flocks on agricultural crops—wheat sprouts, young corn, apples, cherries, and berries. They lurk by the tens of thousands around corporate agriculture facilities and binge on feed in the troughs of cattle and swine, picking out the most nutritious bits and leaving the dross for the farm animals. According to an estimate by Cornell University researchers, in the U.S., starlings cause *eight hundred million dollars* in agricultural damage every year.

But the harm from starlings extends far beyond the agricultural realm. Their swirling skyborne murmurations spiral down into roosts on urban buildings and at suburban park edges, creating noise, odor, and filth. Starling feces in some populations may contain histoplasmosis, a fungus that affects humans and other mammals; any resulting illness is usually so minor that it is undetectable, but the fungus can lead to respiratory infections and, in extreme cases, pneumonia, blindness, and even death.*

In 1960, a Lockheed L-188 Electra serving Eastern Airlines Flight 375 took off from Boston's Logan Airport for Philadelphia and other points south. Seconds after takeoff, the plane collided with a flock of twenty thousand starlings. Hundreds of birds were sucked into the machinery; two of the four engines lost power, and the plane plunged into the sea. Sixty-two people died, including several who were in town for a shoe-sales conference. After the crash, officials tested seasoned pilots on flight simulators to see if any of them could have saved the plane in such a scenario. All failed. In more testing, live starlings were thrown into running engines. It was found that just three or four birds could cause a dangerous power drop. The crash of Flight 375 ushered in a sense of realism that shook the industry and the imagination of a country that, in 1960, was still enamored of air travel. There are multiple stories of starling-airplane collisions, but this is one of the few that resulted in an actual crash or human injury. It remains the worst crash caused by a collision with birds in airline history. After the crash, wildlife-management plans became part of airport construc-

* Droppings would have to build up in an area's soil for at least a couple of years in order for levels of the fungus to become dangerous, and no case of histoplasmosis has been proven to be connected to starling droppings, but the possibility still causes concern and is often mentioned in dissertations on starling damage.

tion and maintenance. Even so, aircraft collisions with starling flocks, or sometimes just the threat of such collisions, result in occasional unscheduled landings, delayed departures, and expensive repairs.

Human deaths are serious, but this crash was a long time ago; most birders know nothing about it and hate starlings anyway. Beyond the costly crop damage, beyond the excrement that collects beneath urban roosts, and beyond the trees full of loud screeching birds, starlings are despised above all else in conservation circles for their ability to outcompete native birds for food and, more important, a limited number of nest sites. Starlings are cavity nesters, and early each spring they will start investigating crevices in buildings, homes, and birdhouses, as well as holes that have been carved into trees and electrical poles by woodpeckers. They compete directly for these prized sites with the other cavity nesters, including chickadees, bluebirds, and swallows.

Self-proclaimed starling vigilantes across the country have taken matters into their own hands. One woman on a Seattle birders' forum proudly catches starlings in Havahart traps, then drowns them; she is downright gleeful when reporting that some of the birds die from overheating in cages left exposed to hours of sun or from fright-inspired little starling heart attacks so she doesn't even have to dunk them. (Havaharts are cages that trap animals without killing them, ironically named in this case.) She urges us all to follow her good example.

On a much wider scale, farmers, government agencies, conservation organizations, and urban businesses fed up with destruction, poop, and noise have for decades been attempting to eradicate, or at least decrease, starling populations. Imaginative but failed efforts have included elaborate traps; explosives; plastic owls; spreading itching powder over foraging areas; irradiating birds with cobalt-60; amplifying starling distress calls; various poisons; toxic chemical sprays; firing pyrotechnics over roosting areas; laying live wire in gathering places to electrocute starlings through their tiny feet; and spraying flocks with a wetting potion that won't dry until the birds freeze to death. The chemical salt DRC-1339, trademarked as Starlicide by Ralston-Purina in the 1960s, kills starlings horribly over a period of days by uremic poisoning. It remains in wide use. In 2015, U.S. government agents killed over a million starlings—more than any other so-called nuisance species.* That number is typical for a given year, but the annual killings have made no dent in starling populations. And they never will. There are simply too many starlings, and they are too good at reproducing and surviving for population-level efforts to be effective. "It's sort of like bailing the ocean with a thimble," lamented the late Richard Dolbeer, who was a well-known wildlife official in Ohio. Instead, we need to address the much more difficult task of thinking ecologically, of creating human-inhabited areas that are less inviting to starlings and that allow native birds to flourish.

Meanwhile, though I am not a starling apologist (I wish

* For comparison, that same year the USDA killed 730 cats, 5,321 white-tailed deer, 61,702 coyotes, and 16,500 double-crested cormorants.

them eradicated from the country as much as anyone—as long as Carmen stays here with me), it is important to consider a few emerging facts about starlings. First, while their populations grew and spread exponentially for decades, they have for the last thirty years or so been stable. In most places, starling counts are no longer increasing. Every species has a carrying capacity—a number of individuals that can thrive in a given place without exhausting the necessary resources. Starling populations seem to have peaked.

Further, at least some of the species' impact on native birdlife may turn out to be more perceived than real. Any observant neighborhood birdwatcher has seen starlings behaving badly to the nicer little birds, so the anecdotal evidence against starlings is strong. Not only do they claim the best cavity nest sites early in the spring, but they sometimes actually invade a nest that is already inhabited by a native bird, throw the bird out in a flurry of snapping bills and slapping wings (occasionally killing the bird; usually not), destroy the bird's eggs, and make themselves at home. But a respected study reveals that rather than giving up, many of these displaced birds simply nest elsewhere. In 2002, researchers at Berkeley completed a years-long survey designed to document the impact of starlings on native birds. To their amazement, they were not able to determine quantifiable harm. Historical population records for the twenty-seven cavity-nesting species believed to be most at risk from starlings were examined from pre-starling times to the present; the species included woodpeckers, kestrels, swallows, flycatchers, and bluebirds. Most of the species' populations showed no decline, not even the red-bellied woodpecker's, a species of most concern because nest usurpation by starlings has long been observed and recorded. Five species in the group showed insignificant declines, and five species' populations actually increased, but none of these changes appeared to be directly linked to the presence of starlings.

I asked lead author Walter Koenig, now at Cornell, how he felt about the study's findings, which he'd known would be unpopular in conservation circles, where the hatred of starlings is an unquestioned—almost cherished—conviction. "I'm not sure I was surprised by the results," he told me, "but I was a tad annoyed." Dr. Koenig studies acorn woodpeckers, and starlings will often usurp the only nests in his study site. He is the last person who would want to exonerate these birds. Still, he cannot claim that his woodpeckers' overall populations are affected. "The bottom line," says Koenig, "is that we know that starlings are quite aggressive and compete for nest cavities with a whole slew of native species. But the evidence that this competition has led to significant population declines is pretty slim, at best." Yet like most ornithologists, he isn't about to go soft on starlings: "I certainly can't say that's changed my attitude toward them; I still don't hesitate to shoot them when I have the chance."

In 1939, a not-yet-famous Rachel Carson penned an essay entitled "How About Citizenship Papers for the Starling?" In it she argued that instead of seeing the bird as an invader, people should accept starlings as a regular species in the native avifauna and give up talk of "invasive" and "nonnative." After all, the bird was here to stay and was, moreover, earning its keep, since starlings feast gluttonously on cutworms, an agricultural menace. (As naturalist George Laycock put it, "Starlings do nothing in moderation.") Carson's notion is echoed in the stance of some modern conservationists, who are turning a corner in the way they think about some invasives. There are many invasive species, like the starling, that are simply ineradicable; instead of spending time and effort worrying about such species, the argument runs, we should accept them as part of the changing modern landscape and move on to issues that we can actually do something about.

Koenig's results notwithstanding, I believe that such thinking leads to a dangerous complacency. Yes, starlings are a permanent part of the urban landscape, and I absolutely do not support harming individual birds (I won't be picking up a shotgun anytime soon). But like Koenig, I am not ready to quit defending native habitats from this invasive species. For one thing, though Koenig's study is a good one, it is not definitive. He recognizes that the historical data he examined may not have been gathered consistently and that if the study had continued, we would likely find that starling competition, alongside native habitat loss from human growth, negatively affects some species—not just the cavity nesters, but also the less aggressive songbirds that feed in areas where starling numbers are high. Of course, all the other impacts of starlings, including those on large and small farms, remain unquestioned. But in terms of conservation, the most significant point to remember is that starlings thrive in areas that are disturbed by human presence, including dense urban environments—places that more sensitive species simply cannot survive in the long term. For now, it seems some birds go elsewhere when their nests are usurped by starlings. But as human sprawl continues, good habitat areas are getting smaller and smaller and may someday disappear altogether. What happens when there *is* no "elsewhere"? Do we shrug our shoulders and accept that we have created a world in which only starlings and a few other robust species can manage to thrive?

Regarding starlings, we can all share responsibility for keeping their numbers to a minimum by covering tubes and other openings on our homes that provide possible nest sites; putting up nest boxes with holes suitable for chickadees, swallows, and bluebirds but too small for starlings; and removing the starling nests and eggs that do appear. But our task is not simply to get rid of starlings. We need to design human landscapes that are hospitable to more species of native birds. This means less grass and more trees. We need to lobby for the creation and protection of woodland parks and forests on large and small scales.

Recent studies on the presence of trees show us two beautiful and related facts: that even a few trees in urban neighborhoods will increase the diversity of bird species, and that people who live near trees are healthier—both mentally and physically—than those who don't. A treed landscape benefits birds and humans together. In addition to starlings, some native birds (robins, flickers, and of course crows) seem to manage well in suburban areas with huge expanses of grass. But yards with trees—any trees at all—attract more varieties of native birds, as if by magic. Wrens, chickadees, tanagers, woodland thrushes, woodpeckers of all kinds. It is so simple for all of us to take part in the re-wilding of the places we live every day, to increase beauty, and wilderness, and wildness, even on the smallest scales.

In 1939, when Rachel Carson wrote her essay suggesting people learn to accept starlings, there were far fewer of them around. It is difficult to imagine that Carson, an early ecologist and a lover of birds, would have maintained this position in light of the starlings' increase and impact in the decades to follow (and while backyard starlings might actually benefit our gardens, in agricultural areas, they do more harm than good, no matter how many cutworms they eat). Still, it is thought-provoking to ponder this defense of the country's most despised bird coming from one of its most revered nature writers and defenders.

I do know that Carson would have loved meeting Carmen—it was always a delight for her to interact with individual wild animals, and she was enamored of birds. It would have enchanted her to have Carmen light on her shoulder, poop on her pale blue cashmere sweater, and peck at her midcentury clip-on earrings. Wild starlings fascinated her, too, and her field books contained meticulous notes and sketches of their morphology, behavior, and unique foraging habits.

Like anyone who has spent time observing starlings, Carson perceived that wild starlings work hard for their food. Most birds that eat grubs and worms search for them visually or peck at the earth with a closed bill to seek out nutritious treats. Starlings actually *create holes* in the earth by poking their closed bills into the ground, then using their extra-strong mandibular abductor muscles—the muscles that open the bill—to excavate a hole in which to search for wormy prey. We can discover where starlings have been feasting on our lawns and in our parks by the presence of these gape holes. Starling bodies are built for this; their squat, strong legs keep them close to the ground, providing leverage and, of course, giving starlings their characteristic waddle.

This unique behavior is one of the things that helped starlings take over human-inhabited places. Wherever humans go, we spread grass in our wake—expansive urban parks, suburban lawns, golf courses, graveyards—and grass is the ideal substrate for the starling's gape-foraging technique. But living with Carmen, I've discovered that the distinctive gaping behavior is related not just to how starlings eat, but also to how they *learn.** Most birds explore their world by pecking at it; starlings gather information by gaping at

* Ethologist Konrad Lorenz, who also raised a starling, termed this bill motion *yawning*. In spite of the risk that it might be confused with baby-bird food begging, I prefer to call it gaping. After all, yawning is equally confusing, as the birds are not sleepy when they are exploring this way.

things. If Carmen encounters something new or interesting, even something that is clearly not food, she doesn't poke at it—she attempts to *open it,* placing the point of her closed bill on the spot she wants to investigate, then opening her bill quick and wide, over and over.

When she started to exhibit this behavior at just five weeks old, I began paying more careful attention to the other juvenile starlings in the neighborhood, those just Carmen's age, and realized that though they weren't feeding themselves yet, they were using this adult feeding action to explore their world. Even more intriguing, I observed that adult starlings consistently do this too—explore objects that they likely do not recognize as food sources by this peck-then-open-bill gaping behavior. It's fascinating to watch.

This morning Carmen flew out of her aviary and onto my arm, where she promptly began to explore the folds of my sleeve, investigating every little crease and crinkle, then moving on to the spaces between my fingers and opening them, one by one. I like to hold my fist tightly closed to get her riled up and give her something to do. She raises her hackles and chirps crossly as she pries at my knuckles. It's so much fun to pester her. She attempts to make gape holes in flat pieces of paper, the spaces in a woven rug, locks of hair—anything, everything.*

---

* In this, I realize she is like many creatures, including humans, who use their mode of eating to explore and learn about the world. Baby humans put things in their mouths. So do baby bears. Parrots learn with their tongues (they cannot learn to count objects by looking at dots on cards, as pigeons can, but they *can* count if they are allowed to explore raised dots with their funny, big parrot tongues). Even we grown apes turn things

I figure that since I snatched a starling from its nest, no matter how unwanted it might have been in the larger scheme of things, I should give that starling as good a life as it can have. The opportunity to investigate the world by opening things with its bill is clearly a key to maintaining all the elements of starling social intelligence: curiosity, involvement, exploration, playfulness, mischief. To keep Carmen's starling-faculties sharp, I provide her with plenty of gaping opportunities throughout the day. One of her favorite treats is applesauce, which I offer in a little dish covered with foil. The thing about starlings is that when they are with you, they don't want to play with toys or eat things on their own; they want to play and eat *with you.* So I hold Carmen's applesauce in my hand while she jumps on my thumb and pokes holes in the foil or lifts the corners to get to the yummy fruit. Then I cover it up, and after a disgruntled *squawk,* she does it again. I hide a grape in my fist so she has to find it through my closed fingers. When she does, she eats the grape by opening it with her gaping motion though the hole on top, where the stem was, rather than pecking at it. I give her toys that she can gape into—origami cranes, sponges, snail shells. I cover the bottom of her aviary with multiple layers of newspaper so she can lift and explore them, opening the folds, making little tents, and crawling beneath them,

---

over, exploring visually, smelling, acquainting ourselves with new objects as we would if we were going to pop them into our mouths. Perhaps some of this is lost in very recent times, when much of what we encounter comes from a flat screen—we see different images, but always with the same texture, fragrance, and weight.

sometimes to my great distress—where has she gone? When I call "Carmen," she peeks out one side, as playful as a puppy.

Carmen gaping for applesauce. *(Photograph by Tom Furtwangler)*

I love introducing this goofy, pretty, smart bird to my friends, knowing that she brings to everyone, as she has to me, a renewed sense of the raw beauty and intelligence in all of life, accessible only when we are able to strip away our preconceptions. "I hate starlings," a guest will tell me. I let Carmen out of her aviary. She lights on my hand, stands in her shining feathers, and sweetly tilts her head. And I ask my guest, "Do you hate *this* starling?"

For perspective, I contemplate Mozart, who lived more than a hundred years before starlings had been purposely introduced anywhere, at a time when the human relationship to nature was a subject of philosophical and literary thought but long before anyone had an inkling that ecology would become a subject of study. Mozart lived in a place where starlings were native and thus valued as a providential part of the natural world, a place where their beauty was admired and their voices were appreciated. Even today, starlings are valued in Europe, where their numbers are declining due to the loss of agricultural lands. In England and other parts of Europe, this decline has even led to worry over the birds' welfare—they are officially listed as a species of concern. Mozart's relationship to Star, and that of the friends he introduced to the bird, were unclouded by a hovering hatred for the species.

The place we are left to inhabit in our thinking about starlings is a complicated one, but one that we are equal to. Carmen and her kin invite us to experience the poetic dissonance and multilayered understanding that is one of the hallmarks of our creative human intelligence. Starlings are shimmering, plain, despised, charming, collectively devastating, individually fascinating. We have the capacity to realize that while a species may be ecologically undesirable, the individuals of the species are just birds. Beautiful, conscious, intelligent in their own right. Innocent. Do I want starlings gone? Erased from the face of North America? Yes, unequivocally. Do I resent them as aggressive invaders? Of course. And do I love them? Their bright minds, their sparkling beauty, their unique consciousness, their wild starling voices? Their feathers, brown from one angle, shining from another? Yes, yes, I do.

"There is another world," Paul Éluard wrote, "but it is in

this one."* One world is marked by a bland forgetfulness, where we do not permit ourselves an openness to the simple, graced beauty that is always with us. The other is marked by attentiveness, aliveness, love. This is the state of wonder, which is commonly treated as a passive phenomenon—a kind of visitation or feeling that overcomes us in the face of something *wondrous*. But the ground of the word, the Old English *wundrian,* is very active, meaning "to be affected by one's own astonishment." We decide, moment to moment, if we will allow ourselves to be affected by the presence of this brighter world in our everyday lives. Certainly we get no encouragement from what Clarissa Pinkola Estés calls the "overculture." It cannot be assessed by the standardized cultural criteria of worth—measures that can be labeled with a sum or a statistic or even, perhaps, a word. Receptivity to wonder is not economically productive, marketable, quantifiable. The rewards, also, stand beyond such calculation. But it is in such receptivity that we discover what draws us, and along with it our originality, our creativity, our soulfulness, our gladness, our art. Mozart found inspiration in the presence of a common bird. For us, too, the song of the world so often rises in places we had not thought to look.

* Sherman Alexie chose this quote for the epigraph to his *Absolutely True Diary of a Part-Time Indian;* he attributed it to W. B. Yeats, and others have done the same, but the origin seems to be French surrealist Paul Éluard.

Four

WHAT THE STARLING SAID

HIieeeiEEeee. Carmen's first word was wobbly and uncertain. But it *was* a word. She'd said "Hi." Right? Hadn't she? Tom and I looked at each other, neither of us wanting to be the one to suggest that a random warbling was an intentionally formed humanlike sound produced by our four-month-old starling chick. But then she said it again, clear as the bright August sky. *HIiiiiiii. Hi. Hi*. Tom and I grabbed hands and jumped up and down. "Hi!" we sang back. "Hi, Carmen!" Our baby had said her first word! We ran up to Carmen's aviary and peered inside. "Hi, Carmen!" She hopped over to see us, turned her head as if to listen, and—as she always would in the future when we tried to get her to talk on request—fell completely silent. But as soon as we returned to the kitchen: *Hiii!* We looked back at her. Was she mocking us? Most likely not. She was just enjoying her new way of expressing herself, one that would become, far more than any of us could have imagined, an integral part of the life of our household.

It is a surprise to most contemporary Americans that starlings can talk, that they are gifted mimics of environmental sounds, other birds, music, and the human voice. In this they surpass crows and ravens and are on par with birds of the parrot family. But the fact of starling mimicry was common knowledge in other times and places. Shakespeare clearly expected his sixteenth-century audiences, a hundred and seventy years before Mozart, to understand the reference in *Henry IV* that inspired Eugene Schieffelin's Central Park starling introductions: "I'll have a starling shall be taught to speak." Most of Shakespeare's audience were not aristocratic, or particularly literary, or even educated, yet if they had not known that starlings could talk, the plot point would have made no sense. And such knowledge is older still. In the first century, the Roman author, military commander, and natural philosopher Pliny the Elder raised starlings for study and recorded that Julius Caesar, among other early statesmen, also kept them and taught them to "parle Greek and Latin." Disastrously for the starlings, it was believed in sixteenth-century Britain that if one used a silver sixpence to slit a starling's tongue, the bird could be taught to speak more impressively.

Soon after the *Hi,* we realized that, although Carmen had formed her first human word, it had not been her first act of mimicry. A couple of weeks earlier she had started making an odd sound, one we hadn't heard from her before and that wasn't in the innate starling repertoire of whistles and squawks that she was perfecting in her spare time. It was a quick up and down, an *EEE-oo-EEE-oo-EE-oo.* Three or four like that, in rapid succession. It sounded to me a little like the song of a common yellow-throated warbler, a bird that Carmen had never heard. Then, just a few nights after the *Hi,* I had poured myself a little glass of chardonnay and was using the Vacu Vin (the plastic pump apparatus that sucks the extra air out of a wine bottle through a rubber stopper) when I heard it: *EE-oo-EE-oo-EE-oo!* I stood, mouth agape, until I recovered enough to yell, "Tom! Tom, Tom, Tom!" He came running. "Listen!" And just as I started to lift the pump, Carmen joined in—Vacu Vin and starling almost indistinguishable. I'm sure this reflects poorly on me somehow; the first repetitive environmental noise my starling learned to mimic was the sound of me hitting the bottle!

Her vocabulary grew quickly over the next several months. Other than the initial *hi* greeting, we chose in the main not to attempt to teach Carmen particular words but to let her vocalizations unfold in tune with her life within our family. Constant favorites are *Hi, Carmen; Hi, honey;* and *C'mere!*—the short phrases we utter every day as we pass by her aviary or when we open the hatch and call her over to land on our shoulders. She is good at the kissy sound humans use when calling an animal—she makes three in a row, just as we do with her. Sometimes, when she's not trying very hard, she sounds shrill and starling-ish when she imitates human words. Sometimes she sounds just like a feathery little person.

Meanwhile, Carmen continued to expand her household-sound repertoire. After the wine vacuum came the coffee grinder. Then the beeping of the microwave oven—pitch perfect. She integrated all of these sounds, as well as *Kiss me!* and *C'mere!* and *Hi, honey,* into her rambling song bouts. She would also often make the sounds individually, just calling them out one at a time and at random, or so I thought at first. It took me far too long—and I feel quite stupid about this—to notice something else, something startling and beautiful.

Carmen imitates the coffee grinder—an unpleasantly loud but accurate *Whirrrraaaaah!*—when I open the jar of coffee beans and pour them into the machine. Every evening I walk into the kitchen to grind the beans for the next morning. There is Carmen, eager to announce, first, the *Ker-klunk* of the coffee-jar lid as I set it on the counter, then *Whiirrraaah!* When I open the microwave door, she immediately interjects her eerily mechanical *Beep! Beep-beep!* And the wine-vacuum sound? When she hears the clink of the bottle.

All of this made me realize that, like these household sounds, the words she imitates are not called out at random. When I come downstairs in the morning, it is not the microwave or the kissing sounds I hear first, but the greeting—*Hi, Carmen! Hi, Carmen! Hi, Carmen!*—the first words I say to her each day. And when I stop to peek into her aviary?

C'mere. All participatory, *anticipatory,* all involved, all cognizant of what is going on aurally in her world and what precipitates what. She is basically saying, *I know what's going on! I am part of this!* I am filled with wonder. And with questions.

Meredith West is an ethologist emeritus at Indiana University and the lead author of an iconic 1990 *Scientific American* paper that explored the subject of Mozart's starling and the relational capacities of starling pets. It was West's paper that confirmed for me that the story I'd heard about Mozart's pet was indeed based in fact, and her work brought to light the details of Mozart's notebook for an English-speaking audience. She also kept starlings in her home so she could understand how Mozart fell for such a bird. But the heart of West's work is really the aural complexity of starling vocalizations and the character of communication possible between starlings and humans. According to West, the vocal interaction we experience with Carmen is typical; West refers to it as a kind of social sonar or echolocation: starlings toss out sounds to see what comes back to them. Where for a bat the reward of echolocation is a meal of mosquitoes, for a starling the reward appears to be a sense of comfort in belonging. Carmen's verbal and aural participation is a way of locating herself in her surroundings, which for a starling are not just physical but also social. Response from the world around her is essential.

In 1983, Meredith West and her colleagues carried out a pioneering study on starling mimicry. Seven starlings were captured as five-day-old nestlings, just like Carmen; four were females and three were males. West's chicks were all hand-raised together in the lab until they were about thirty days old, at which point they were split into three groups. Each group went to live in a household with human caretakers, but all under different circumstances. A male and a female went to live in a home where, like Carmen, they were raised with constant human contact and treated as part of the family; they were often let out to fly free, were fed by hand, and were included in group conversations. They were sung to and whistled to. There was no attempt to teach the birds particular words or songs or to mimic any sounds that the starlings made except when doing so came naturally. Another male and female went to a home where the humans didn't interact with them at all except for what little was necessary to keep them fed and their enclosures clean.

The last three birds went to live together on a screened porch in the same home as the first pair of birds, those that were living as part of a family. These screened-off birds could hear everything that the inside birds heard—the human voices, the singing, the vacuum cleaner—but people in the home didn't approach them, talk to them as individuals, or handle them.

It turns out that among starlings that live with humans, Carmen is not alone in her eagerness to join the household's conversation. In West's study, all seven of the birds mimicked natural sounds, mechanical sounds in their environment, and other birds. But only the birds that interacted closely with humans mimicked human words and voices, and only these birds mimicked environmental sounds in the *context* of what their human caretakers were doing. West and the other starling-keepers in her study reported that their birds kept them constantly on their toes, readily mimicking a variety of sounds, most commonly greetings and good-byes (*Hi, Good morning, Hey, buddy, Night-night, Go to your cage*); attributions (*Silly bird, You're so pretty, See you soon, baboon, You're kidding!*); conversational fragments (*Whatcha doing, Okay, I guess so, This is Mrs. Struthers calling*); and household noises (cat meowing, dog barking, door squeaking, keys rattling, dishes clinking). Many people who live with starlings report becoming self-conscious about tics they never knew they had until their starlings mimicked them back: sighing, coughing, sniffing, tongue-clicking, odd little laughs. In Meredith West's academic household, a resident starling would perch on the professor's shoulder and mutter, *Basic research, it's true, I guess that's right,* and when someone else walked into the room, the bird would announce, *I have a question!* The implication is that mimicry has a rich and complex social aspect—that it's valuable and useful for starlings to connect aurally with those they are most closely bonded to, whether that is another starling or a human.

West and her team also found that the family-raised birds imitated the cadence of human speech whether they were forming recognizable words or not, just like a human baby starts to sound like it is making words before it actually is—the bird version of baby talk. Carmen does this all the time. She will be rattling away, whistling and gurgling like a wild starling, and I'll look at her and say, "What is all this noise?" She'll stop the starling talk, tilt her head, and say with a perfect American English lilt, *Why, it's nothing. Say, that soup on the stove smells delicious. Why don't you let me out and we'll have a nice game of backgammon while it simmers.* She doesn't produce the actual words, of course, just the up-and-down cadence and intonation, almost exactly like my own.

It's the same as a person imitating French or Chinese or some other language that they don't actually speak; it is easy to guess what language is being imitated even though the words are nonsense. When my family was over this holiday season, they kept telling me, "Carmen said, 'Merry Christmas!' I heard her!" She didn't, but you could imagine it. She was just baby-talking household English. (And let me offer a piece of advice: *never* teach a starling to say "Merry Christmas" unless you want to hear it all year long.) Birds raised within earshot of humans but without close contact do not behave this way. Carmen's vocalizations are *relational*, a kind of conversation. They are her way of *being with us.*

It's kind of unnerving, actually, to be sitting alone in the living room on a dark rainy night, cuddled down by the fire with a book, and hear a little voice from the next room say with perfect inflection, *C'mere!* She sounds so hopeful, I can't help getting up and walking over. "Here I am." Carmen jumps to the front perch, puts her face to the wire, and we touch, nose to bill. "Go to sleep," I tell her, and turn out all the lights.*

All of this makes me ponder the aural relationship between Mozart and Star. By Mozart's time, the keeping and training of birds with good singing voices had become common among the European middle and upper classes, aristocracy, and even royalty. In the early 1700s, royal Parisian "supervisor of the woods" Hervieux de Chanteloup had a second job description: "guardian of the princess's canary birds." Most in demand were the bird species with vocal agility: canaries, bullfinches, linnets, East Indian nightingales (actually a kind of mynah), and starlings.

Little eight-holed bird flutes called flageolets became popular among those with the leisure to train their birds in song, a fashionable hobby. (Mozart would absolutely have known about this trend, but it is unlikely that, as a professional composer with a houseful of fine instruments, he would have deigned to purchase one of the popular flutes.) In 1717, *The Bird Fancyer's Delight* was published in London,

* I would have her come out and read with me, but like many birds, Carmen acquires a kind of confused agitation as the sun goes down. Keepers of budgies and other pet birds will recognize the behavior; if she's not inside her aviary when darkness falls, she gets disoriented, bashing into walls and windows that she navigates readily during the day, and she is easily startled. We make sure she is closed safely in her aviary by sunset.

a slender book of tunes "properly compos'd within the compass and faculty of each bird," with a few tunes for each of eleven species, including three for the starling.

The pieces for starlings in *The Bird Fancyer's Delight* are decidedly un-starling-esque, composed in a straight G major or F major and rolling along prettily and predictably, with no flights of starling fancy whatever. I have performed them for Carmen on various instruments—violin, piano, harp—and asked my daughter to play them on Carmen's favorite instrument, the cello. Carmen does what she always does when we play music for her: she tilts her head attentively, and curiously. She is, as ever, a good listener. When music she loves comes on the stereo or is performed live on one of the household instruments, she sings along exuberantly. But with these tunes, she offers nothing more than a polite little squeak here and there, and perhaps a perfunctory whistle and wing fluff. She is unimpressed.

I can't imagine that eighteenth-century starlings would have displayed any more interest in *The Bird Fancyer's Delight* than Carmen does, but bullfinches proved to be apt pupils of the tunes and learned their songs readily. Bullfinches are fat, handsome little birds topped with oil-black caps and gray backs that are set off against unusual persimmon-colored breasts. In Germany, bullfinches were captured in groups and taught to sing for show and for sale. The popular method of professional bird trainers was to keep the young birds in complete silence and darkness, often without food, so their senses would be formed by nothing but the tutored tunes. Some trainers even blinded their birds with hot metal spikes. This kind of "music by torture," as composer David Rothenberg called it in his excellent *Why Birds Sing,* was often successful with the gifted bullfinches. William Cowper's 1788 poem "On the Death of Mrs. Throckmorton's Bullfinch" attests:

And though by nature mute,
Or only with a whistle blessed,
Well-taught, he all the sounds expressed
Of flageolet or flute.

Those that survived the abuse and sang well commanded high prices in the shops. Star, worth just a few kreuzer with his bits of mimicked motifs, would have been considered a musical hack by comparison. At least at first glance.

Among the birds commonly kept for song, only the starlings and mynahs were true mimics. It's an important distinction. Most songbirds possess an inherited ability to learn the songs unique to their own species, usually sung by the male, most often during breeding season. This spring song is a proclamation of breeding readiness, and a strong, exuberant song helps to attract a mate and declare and defend a nesting territory. If you take a walk in the woods with a good birder in the right season, she will be able to announce the identity of the birds singing high and invisibly in the leafy trees based solely on their song. Most passerines learn their species' song when they are young, so they are ready for the breeding season the year after their birth. In the autumn we can hear the young males of vocal species like house finches or white-crowned and white-throated sparrows singing wobbly versions of their songs—beginning practice for the spring. Sweet sounds. And for most of these birds, there is a learning window; if they do not learn their species' songs in the first year or so of life, they are unlikely to learn them at all. Some birds raised in captivity will be able to sing a version of their species' song, usually not a very good one, but many need the live tutoring they receive in the wild from older males. And if captive birds are introduced to another song instead of their own species' song during this sensitive song-learning window, some will learn that song instead, which means that for some species, the song is learned rather than innate. These birds have an inherited tendency to learn the song they are exposed to most often when very young, just as the trained pet bullfinches did, as long as it is in the vocal range of their usual species' song.

This is very different from the true mimicry we see in starlings. Mimics do sing a song unique to their species (in the case of the starling, this includes a long series of teakettle whistles, clattering, and shrieks that many are reluctant to term a *song*). But true mimics like starlings do something else too: they imitate sounds from their environment—novel and improbable sounds that lie far outside of the usual explanation for birdsong. They appear to select these sounds at will; we can attempt to teach starlings sounds and tunes, and they will turn their nose up at some and latch onto others. They are mysteriously discerning and certainly have no regard for what we want them to mimic. Meredith West thought her birds had failed to learn anything from words and songs she'd played them on cassette tapes until she discovered one of them mimicking the *hissss* of the recording tape between phrases.

This kind of true mimicry is uncommon. Passerine mimics include the starlings and mynahs, the corvids (jays, ravens, magpies, crows), and mockingbirds. Outside of the passerines, parrots are the most famous and gifted mimics. All these birds will imitate other birds and animals, environmental sounds, musical motifs, and human voices. While a young bullfinch can learn to sing a song in its impressive native range, unlike a mimic, it cannot imitate random and novel phrases.

Mimicry indicates a sophistication in relationship to sound and involves a plasticity of behavior and consciousness that goes far beyond the instinctive. New research at Duke University suggests that the brains of parrots display a unique genetic pattern. The area of the brain associated with vocalization is surrounded by another layer, or shell, of ultraspecialized neurons for sound recognition and creation—a "song system within a song system," as the researchers put it. This may be true for the brains of non-parrot mimics as well.

Humans, ever self-absorbed, homed in on the capacity of certain birds to imitate our own speech right away, and it became one of the first areas of avian vocal research. Pliny the Elder kept magpies alongside his starlings. He noticed how eager the birds were to learn new words and how, once fixed on a word, they would work and fuss over it until it was perfected. Carmen certainly does this. One of the first phrases she chose to mimic among those oft heard in our home was *Hi, honey!* The word *honey* came out rickety and birdish and strange for days, but she'd turn it over and over on her tongue. And when she had it? She proclaimed it loudly and, to all appearances, even joyfully—not shyly, as when she was still perfecting the sound. Pliny declared that a magpie who has trouble learning a word will suffer such angst over his failure that he will die of it! He mused wonderfully: "They get fond of uttering particular words, and not only learn them but love them and secretly ponder them with careful reflection, not concealing their engrossment.... It is an established fact that if the difficulty of a word beats them this causes their death." Secretly pondering? Perhaps. But Carmen, I hope, is not so angst-ridden.

In spite of Pliny's early start on the subject, and even after centuries of study, mimicry in birds is little understood. No single theory applies for all birds, and there are elements of mimicry that defy explanation altogether. The classic theory with starlings is that males use mimicry in the main to impress females. As a female starling, Carmen is not supposed to be a gifted mimic or songster. But living with a female starling, I can heartily attest that they create many of the same song elements that males do, and obviously they can be capable mimics. I was shocked and alarmed, though, when in April of Carmen's first year of life she stopped vocalizing. Completely. She didn't just stop the incessant whistles and the *Hi, Carmen*s and *C'mere*s. It was every single little peep. I freaked out. I coddled her, sang to her, exhorted Claire to set up her cello next to Carmen's aviary and play her favorite Bach suites, with which she had always loved to sing along. Nothing. While I fretted about how I had failed as a starling-keeper, I developed two theories. First, I thought that as a social bird, Carmen might be depressed in her life without starling companionship, that maybe her vocal and social skills had slowly been withering away and I was too dull to notice until she stopped talking altogether. I continued to spend extra time with her, added more mirrors to her habitat so she could keep company with her own image when alone, played tapes of starling vocalizations, and left her listening to her favorite bluegrass music when I had to leave. Nothing. Not one little tweet. She was as friendly and tame as ever, but she didn't make a sound for nearly three whole months.

During this time, I worked out my second theory. This was the season that, were she not living in a house, Carmen would have been nesting. Even though she had no mate, no nest, not even an artificial nest box, and even though she did not lay eggs, she might still have been overcome with the biological imperative to keep her nonexistent nest and nestlings safe by remaining absolutely quiet. Her mate, if she'd had one, would accomplish the necessary scolding and chest puffing required to guard the nest from interlopers or predators. Her job was to be silent and still.

I granted that this was a more likely hypothesis, but it still seemed improbable. She was so far removed from the nesting process, why would this one behavioral imperative take over? And yet, why should it be any different from the other physiological changes associated with the seasons that she *was* displaying? Her legs became lighter pink, her bill changed from dark to bright golden yellow—both signs of breeding readiness. Perhaps this silence was simply seasonal? I did what I always do when Carmen's behavior baffles me: I threw binoculars around my neck, grabbed a notebook, and headed out into the field—in this case, my urban neighborhood—to see what the wild starlings were up to. Sure enough, the male birds were in full-voiced territorial glory, and the females were hidden in their nests, silent as stones. This gave me modest hope.

Finally, in July, just when I was beginning to despair of ever hearing her voice again, I padded down to make my morning coffee and heard *Hi, honey!* "Carmen!" I ran over to her aviary, but she was nonchalant, as if nothing had ever been amiss. Then she broke into a long, discordant whistle, and over the next few days she returned to her full, spirited voice. This was just when her brood of nonexistent chicks would be fledged and gone, giving her no further need to be secretive and hiding.

If you look into starling-vocalization research, you'll find papers with titles like "Temporal and Sequential Organization of Song Bouts in the Starling." This one describes the function of mimicry as purely mate attraction and makes no mention whatever of female voices; every single bird studied was male. Or there is "Song Acquisition in *Sturnus vulgaris*: A Comparison of the Songs of Live-Tutored, Tape-Tutored, Untutored, and Wild-Caught Males." Males. The study methods make sense; in most passerine species, while females make various chips and calls, it *is* the males that do all the true singing, so naturally the study of singing starlings has focused on males during their peak singing time—spring and summer, the months that, as Carmen taught me, female starlings go silent. No wonder the females' vocal capacity is not just underestimated, but almost entirely unknown.

I spoke to Meredith West about the skewed view of female starling voices. Her study was unique in that more than half the birds involved were female. This was just the luck of the draw—it's hard to sex baby starlings, and they took what they got. In the end, this was fortunate. While it wasn't something West set out to prove, her work uncloaked the extraordinary vocal prowess of female starlings. She agreed with my finding that female starlings could be as vocal and gifted mimics as males, especially when they are in a social group (like a family) with whom they want to interact.

Both male and female starlings invite researchers to expand on the classic explanation for mimicry. Sure, there is a role in pair-bonding, and males do seem to take the lead in using mimicry to attract a mate. But once the pair bond is secure, mimicry on both sides appears to be a way of maintaining intimacy between mates. Through the seasons of the year, mimicry continues in both sexes, even though males and females often split into separate flocks; it's a form of connection and belonging among flock-mates, of environmental awareness and participation. I am certain that there is more going on with both male and female starling communication and consciousness than we realize, but so much of this understanding cannot be learned in the lab, or even in the field, where we experience the habits of starlings in fits and starts. It can be learned only by the rare privilege of living in constant contact with a wild bird.

Carmen fluffed after a bath and surveying her world. *(Photograph by Tom Furtwangler)*

Just recently Carmen added a new trick: meowing like the cat. Delilah appeared entirely unamused. Tom, Claire, and I all looked at one another, and Claire sighed. "It was bound to happen." From the next room, we can't tell the two apart. Delilah's food bowl is near Carmen's aviary. Unfortunately, this means that the meow Carmen has learned is not some sweet, purry, happy meow, but a crabby-cat *Feed me!* meow, the one our fat Delilah makes while circling her bowl, announcing her firm belief that it is mealtime. As usual with a new sound, Carmen worked on it, perfected it, and then mixed it into her repertoire—those bouts of singing where she throws all the mimicked sounds she knows into one long string. She would, as she did with other sounds, single out the *meow* and call it at seemingly random moments. And, also as usual, she did something else that took us way too long to comprehend.

Tom noticed it first. He walked into the kitchen and Carmen looked at him and said, *Hi, Carmen!* Then Delilah walked in, and Carmen looked at her and said, *Meow!* At first we thought it had to be chance, but it has happened too many times, too consistently. Carmen greets us, human and cat, each in our own language.

Carmen has learned that *Hi, Carmen* is a greeting, and she uses it as one. But what about that *C'mere*? This phrase, in the context of Carmen's life, usually means we are going to hang out together. I open her cage, say, "Come here," and hold out my hand. What about those times that I am in the next room, and I hear her call, *C'mere?* Is it a signal of desire? Does she *want* me to come there? I would not suggest that she has any sense of the grammatical structure of the words, or even that she recognizes them *as words,* as compared to environmental sounds. But given what else she has taught me about the way that starlings understand context, it is altogether possible that she has learned the cause-and-effect relationship between the sound of *C'mere* and the result—my presence, her favorite thing.

I realize that this might sound far-fetched. We are talking about a bird, and such complexity of consciousness exceeds what we typically imagine birds to be capable of. But really, what I am suggesting is analogous to the dog who carries its leash over to its person and looks up expectantly. Leash equals hope of a walk. Why couldn't Carmen's *C'mere* not equal the hope of Lyanda coming over to the aviary? This question presses the boundaries of what we know about avian consciousness and human-bird communication, yes, but it is past time for such pressing. For so long, birdsong has been considered a function of breeding and territoriality. But the earth and its beings are extravagantly wild, full of unexpected wonders. It is time to turn from our textbooks and listen to the birds themselves.

Carmen's influence on my never-ending life pilgrimage in the natural world is profound. I find that when I walk into the world these days, I cannot help but say "Hi" to the starlings I see, and all the other birds too. I am so accustomed to Carmen's friendliness, I almost expect that one of them will fly to my shoulder and say *Hi* back. Writing on her blog *Myth and Moor,* author, artist, and folklorist Terri Windling reminds us, "Many an old story begins with the words, 'Long ago, when animals could speak…,' invoking a time when the boundary lines between the human and the animal worlds were less clearly drawn than they are today, and more easily crossed." Perhaps the corollary would be just as good an opening for a tale; not "Long ago, when animals could speak," but "Long ago, *when people could listen*." I look back to my naive early notion that I would obtain a starling to study in support of my own ideas, the story I thought I wanted to tell. But when I manage to hush my own voice and just listen, I discover that Carmen has become not just part of the story, but the *storyteller,* whispering in my ear, telling me what needs to be written, to be spoken, to be sung into the world.

Five

THE STARLING OF VIENNA

When Carmen was young—able to fly but still fat and gray-feathered—she used to explore our house courageously. She'd fly around every room and liked to sit high atop bookshelves, or lamps, or chandeliers, surveying the scene. Nowadays she will come when she's called—or sometimes she will, anyway. I'll hold out an arm and call her name, and though she will ponder for a moment whether or not to do as she's asked, she will most often decide in my favor. Baby Carmen did no such thing.

As songbirds feather out, bits of downiness remain here and there among the new feathers, and the last of this to fall away is the one little tuft over each eye. A bit of downy fluff sticks out there, like an old man's eyebrows, giving the young birds a half-adorable, half-grumpy look. Carmen, with her brows fluttering, would gaze down at me from the top of some high cupboard when it was time for her to go in for the night. I'd hold up my arm hopelessly. "Come on, Carmen! Come on down!" She'd just look sweetly about, as young birds do. Finally I would pull over a stool and risk my life climbing sock-footed onto the counter. When I held my hand close to Carmen, she'd happily hop on.

Our house was built in the 1920s. Its most prominent features are wide fir moldings and high airy ceilings. The floor plan is circular; kitchen to hall to living room to dining room to mudroom, and back to kitchen again, with wide archways between each room. When Carmen was just beginning to fly, I'd help her practice by running in circles around the main floor. She would follow me, flying behind like a little kite. When she (or I) got tired, we'd take a break and she'd plop on my shoulder for a rest. Then I'd start running again, and she'd leap back into flight.

She used to fearlessly follow me everywhere. During my morning shower, she would flit from the shower-curtain rod to the top of my head. When I shampooed my hair, she would gape through my locks to explore the suds, so I put a stop to this practice, worried that she was ingesting soap.

Grown-up Carmen does none of these things. Young birds are sweetly naive and open to exploring their world without caution. This is part of why mortality is so high in the first months of a bird's life. Adult birds grow wise and more wary, for good reason.

The longer Carmen lives with us, the more she settles into a routine. She is comfortable and unafraid in the kitchen, dining room, my study, and of course her aviary. If I wander elsewhere in the house, she becomes visibly nervous and will stick steadfastly to my shoulder. Overall, though, I don't see her as more fearful in life, but more tranquil. I love to see that she has, in her way, made herself *at home*—made her own home within our larger one, with her own paths and ways and routines and places of comfort.

In my mind, Carmen's unfolding story in our household—both physical and aural—mingles constantly with the story of Mozart and Star. I live with not just a wild bird, but *Mozart's bird*. Every day I pick up my pen and smile, just as Mozart did, at this starling's iridescence, her wildness, her chattering. I transport myself by the power of visualization to the Mozarts' home, where Star, like Carmen, would have settled in his own way into their family routine. Even across the distance of time and culture, Mozart's life with Star and mine with Carmen surely share a great deal. Yes, Mozart was a musical genius. But in the bare practical outlines, we are two writers, sitting at our desks, with starlings on our shoulders.

During the very first days that I started to wonder over the story of Mozart and his starling, I began poring over guidebooks to Vienna and Salzburg and treatises on the life and culture of these places. I tried to re-create the streets that Mozart walked and the rooms where he lived with Star in my mind's eye. I am armed with a tenacious conviction that somehow the presence of the people who live in a home resides in the atmosphere of the walls forever. Although Carmen could teach me a great deal about how anyone, including Mozart, would have lived with such a unique bird, I knew that Mozart's rooms held even more secrets. Like so many Mozart pilgrims before me, I packed my bags for Vienna. First, I would find the perfect Sacher torte. Then I would wander the streets of the city and the halls of Mozart's home.

It was September 29, 1784, when the young Mozart couple moved from their cramped quarters on the Graben to sizable apartments on Domgasse, just around the corner from the great spire of St. Stephen's Cathedral, Vienna's beloved landmark, visible for miles around. In the carriage, Star likely rode in a small cage alongside the Mozarts' swaddled son, Carl Thomas, just nine days old. The weather was fair, and in spite of the trouble of relocating with a newborn, Mozart and Constanze were surely in good spirits. Their boy was rosy, and they were moving to gorgeous third-floor apartments in a desirable neighborhood. Besides, they had to travel only a few blocks, just a ten-minute walk from their previous home, perhaps a few minutes longer in laden carriages. Though the Mozarts moved frequently during their life together in Vienna—fourteen apartments in just ten years—the home at Domgasse is the only residence that still exists.*

* The physical exterior of the Graben apartment building still stands, but the interior has been extensively remodeled as a hotel, so the rooms are nothing like they were during Mozart's life.

Anyone seeking to understand this period of Viennese history and culture is indebted to the rare, on-the-ground writings of cultural commentator Johann Pezzl, who between 1786 and 1790 compiled some of the only extant eyewitness cultural commentary on the place and time. I studied his *Vienna Sketches* obsessively in the weeks before my journey. Pezzl reports that the third floor (what Americans would call the second floor, considered the third floor in Vienna, where the basement is counted as floor one), although the most expensive, was by far the most coveted—another reason the Mozarts were thrilled with their move. The ground-level floor and the one above it were undesirable because of their nearness to "dust from the street, the smells of stables and sewers, and the noise of wheeled vehicles passing outside." But once you got above the third floor, rent became cheaper again, since, though the air and the views were finer, "it is hard work carrying the necessities of life, wood, water, etc. to these heavenly heights, and while the number of steps brings a reduction in rent, it increased the price to be paid for delivery of goods carried up 150 steps ten times a day." Thus, "in the top floors of city buildings, in garrets and in attics, nestle the poorest type of tailors, copyists, gilders, music copyists, wood-carvers, painters, and so-on." Pezzl's view is as political as it is colorful: "These attic floors are often crawling with hordes of children, whose numbers and constant requirements often worry the poor father to the same extent as the rich and distinguished man living on the second floor below has his worries about being able to find a sole heir for his family."

In the fresh modern air, it is difficult to imagine the horrific dust in the lives of all Mozart-era Viennese, which Pezzl describes as the greatest of plagues. "It is the dried-out dust of chalk and gravel, it irritates the eyes and causes all sorts of lung complaints. Servants, runners, hairdressers, coachmen, soldiers, etc., who have to be out on the streets a great deal, often die of pneumonia phthisis, consumption, chest infections, etc.... The worst situation occurs when, after several warm days, a strong wind springs up... the dust penetrates mouth, nose and ears... and one's eyes weep." The Mozarts were better situated in these apartments than we realize at first glance from our modern, dustless vantage point. Above the dust, below the labor.

Der Mehlmarkt in Wein, the center of Vienna in Mozart's time. *(Bernardo Bellotto, 1760)*

I visited Vienna and Salzburg in the autumn, just after the tourist season, when crowds were smaller and prices were a bit lower, but the weather would be—if all went well—still fine. Fortune smiled; my days in Austria were warm and idyllic. Even today, overlaid with tourist-trade Mozartian tchotchkes and chocolates, Vienna feels almost enchanted. The musical soul of the city is palpable and true.

I like to fancy myself an easy traveler, open to the serendipity of a journey, but on the streets of Vienna, I found myself, as I usually do in such scenarios, clutching my *Rough Guide* like Lucy Honeychurch with her *Baedeker.* The map within led me to the Mozarts' Domgasse apartments, now the Mozarthaus museum. Though Wolfgang and Constanze loved these rooms, where they lived for almost three years, they had to leave in 1787 because the rent was just too high. In the decades that followed, there were various leaseholders, and, though no one knows exactly when it happened, at some point the rooms were subdivided into three smaller living spaces.

In 1941, Nazi politicians supervised the acquisition of the lease to one of these smaller apartments and created a modest museum to commemorate the 150th anniversary of Mozart's death for the German Reich's Mozart Week in November of that year. Historian Erik Levi, in his book *Mozart and the Nazis: How the Third Reich Abused a Cultural Icon,* explains that the Nazis were particularly keen to exploit this anniversary, seeing the "universal accessibility" of Mozart's music as an "ideal emotional link between the home and the war front." Ultimately, Levi claims, Mozart was more difficult than many other popular composers for the Nazis to colonize, as his philosophical and moral outlooks, overtly at odds with their agenda, were so well recorded in his many letters.

It wasn't until 1976 that the City of Vienna acquired the lease to all of the Domgasse rooms. The original layout was restored, and a simple museum was set up. In the mid-1990s, the Wien Museum organization hoped to attract more visitors by reinvigorating the presentation of Mozart's former residence. The rooms were primped, and scrubbed, and instilled with a modern curatorial philosophy. Even so, the review in the *Rough Guide* was unenthusiastic. I could not wait to see this house and read over the pages dozens of times before my trip searching for a glimmer of hope, but I found none. This is what the *Rough Guide* reports:

> Sadly...despite all the history, the museum is a bit disappointing. A lift whisks you to the third floor, so you have to wade through two floors of manuscript facsimiles and portraits before you reach the apartment itself. Only one of Mozart's rooms actually retains the original decor of marble and stucco...and there are none of Mozart's personal effects and precious little atmosphere.

My other favorite guide, Rick Steves, did nothing to revive my dwindling hope for this museum pilgrimage, which I had dreamed would be the high point of my Viennese journey. Note that both reviews begin with the same word:

> Sadly, visiting this museum is like reading a book standing up—rather than turning pages, you walk from room to room. There are almost no real artifacts.

Dear me. But I had traveled across the globe to research these very quarters, to see how Wolfgang and Constanze made their life in Vienna and how Star might have fit into it. I lifted my chin and stepped off the narrow cobbled street, prepared to be underwhelmed.

Instead, I was transformed. Certainly the entrance is a bit sad. After a promising approach through a marble archway, the place that horse-drawn carriages would have deposited Mozartian visitors, the new reception area evokes the lobby of a miniature strip-mall Cineplex. There is flat gray carpet, a coatroom lined with cold metal hangers (complete with a stern overseer demanding your wrap and bag), and a discouraging coffee room with Ikea-style chairs and a vending machine. Not an inch of charm in sight. But soon you take the stairs up and up, leaving the modern carpet behind, until you are peering over a rail into the open courtyard and then facing the servants' door into the first room of the Mozart home. The rooms and hallways have indeed been scoured, painted, and made safe enough for toddlers of overprotective modern tourists to wander the stairwells without getting their little heads stuck between the iron rails. But the apartments themselves, the rooms in which the family lived their daily round, are otherwise little changed since Mozart's time.

The critique of the museum in the *Rough* and other guidebooks comes, I think, from the reasoning behind its curation. The accepted standard when creating museums around the residences of famous artists and composers has been to do up the rooms in the fashion of the time, to create a memorial to the imagined life of the subject. If the original furniture of the inhabitant is not extant (Mozart's is not), then similar period furniture or good reproductions are secured and placed in the rooms as the curator surmises they might have been arranged by the famous occupants. This is what we have come to expect and enjoy. The more progressive curation philosophy at Mozarthaus revolves around an admission that we don't know for certain what function each room served for the Mozart family. Instead of imposing their own, possibly erroneous vision on an unsuspecting public, the curators take the refreshingly modern approach of inviting visitors to join them in *imagining* how the rooms might have been used. They speculate in the commentary that is tacked to the walls, but the suggestions are full of accompanying questions marks—"Question Mark by Way of Invitation," they call it: *Study (?); Bedroom (?).* Instead of certainty, we are offered a moment of creative vision. *Here, visitor, you stand in this room, in the rustle of air carrying voices of the past, arias risen to the rafters. You are here. What do you think?*

In my opinion this is not at all boring. It is challenging, beautiful, and alive. Wandering the rooms for hours, I felt that my own imagination was carried away in just the right fashion—with an air of possibility that would not have been attainable in a more traditional exhibit. For surely it is the apartment itself and the whispering ghosts who reside here that are the gem of the museum. And I realized with delight that the curator's approach mirrors the path of the story I am weaving around Mozart and his starling, which comes to life in these rooms but must remain riddled with essential question marks.

In each room there is placed just one period artifact suggestive of life in the apartments. These are all eighteenth-century pieces, similar to the Mozarts' own, though none of them are original to the household. A table in the probable parlor. A single fork in the kitchen. Meanwhile, an artist has created miniature ivory-colored reproductions of all the furniture known to have belonged to the Mozarts. These live in a model of the house in the first room, the servants' quarters, and we are invited to look inside, to arrange the furniture in our mind's eye. The design of the furniture is mainly rococo and baroque, as middle-class families like Mozart's rarely had a house furnished entirely in the neoclassical style of the time.

The museum curators offer another playful bit of whimsy on the wall behind the furniture-filled dollhouse. A small shelf displays a set of tiny statues representing the key characters in the household—Constanze, Mozart, and Carl Thomas, and a crib holding baby Johann Thomas Leopold, who was born at the apartment in October of 1786 and lived just a few weeks. Each figure is formed in red or yellow resin and is about as big as my thumb. To my great surprise and happiness, I found statues not just of the humans of the household but also of the Mozarts' little dog, Gauckerl, and a bird in a cage: Mozart's starling. I leaned over and peered excitedly at the little yellow bird. Admittedly, the statue does not look much like a starling, but who cares? I was thrilled to see that Star was not just my own idiosyncratic obsession but part of this modern telling of Mozart's story.

The professed purpose of the statuettes is to aid us in our imaginative wandering through the apartment. In each room, the figurines of the characters that were likely to have used it are placed in a row on a narrow shelf alongside the curatorial commentary. Thus, Wolfgang and Constanze stand in the parlor. The couple and both boys, who likely slept in the same room with their parents, are in the small bedroom. Gauckerl is pretty much everywhere. And Star? Visions of Carmen's life in my own home swirled through my imagination, and I felt as if this bird figurine leapt from its shelf as a full-size starling and perched on my shoulder. With this unusual guide, I wandered from room to room, asking my question: *Where did you fly?*

I started in the servants' quarters, where cleaning supplies, firewood, and the maid's bed were likely kept; the room is narrow but bright. It is unclear where Mozart's manservant, Joseph, would have slept, though he surely lived with the family. It is possible that he unrolled a mattress in the hallway or kitchen at night and rolled it back up during the day (in this he would have fared better than the other servants and even the family in winter, as his bed would have been nearest the woodstove, the warmest place and only source of heat in the freezing-cold house). This was the Enlightenment, not an episode of *Downton Abbey.* Even in the richest households, a family's relationship with the servants was more casual than it was in those post-Edwardian times, or in the Victorian era, still a hundred years away. The original kitchen was long ago demolished to allow for a central elevator—not for the exhibit, but for the modern residents who inhabit the other apartments. The rest of the building, in fact, still fulfills its original function; several families live upstairs in the four-hundred-year-old rooms above the exhibits.

After you pass through the servants' room (period item: portable wrought-iron candlestick), a small dining room (porcelain fruit dish), and what was likely a sort of guest room (simple wooden chair), the apartment opens out into the largest and sunniest space on the floor, probably the salon. The curators have placed a gaming table from Mozart's time in the center of this room, since a full-size billiards table would have been too large, though Mozart loved the game. This is likely the room where parties, dances, and musical performances were held, with chairs rearranged for guests and musicians to suit the occasion.

The rooms progress in a kind of L shape, and at the far

end of the apartment is the door to a small bedroom with a window that looks out on the cathedral spire. The ceiling is layered with thick, carved, gilded stucco—coils of flowers and fleurs-de-lis and cherubs better suited to an aristocratic mansion than these middle-class rooms. In the 1720s, the apartment was owned by master stucco artist Albert Camesina, who worked for the imperial court, and it was he who added this exquisite ceiling. Stucco workers of Camesina's caliber were not considered simple builders; they were well respected as skilled artisans. (In Mozart's time, the apartment was called Camesina House; later, it was known as Figaro House in honor of the famous opera that was composed under its roof.) In my imagination it was this ceiling that sealed Mozart's delight in these lodgings. He could lie on his back and let the music in his mind and the stresses on his nerves melt into the imagery over his head. There he felt calm, bohemian, and a little rich.

I had been searching the tiny groupings of statuettes expectantly as I explored each room. But after the suggestion of Star's existence at the very beginning and the promise that we would see the figurines throughout the exhibit, there were no further bird statuettes. I queried the docent, who laughed good-naturedly but did not seem to share my wondering over the bird. (Later, I called the historical society. Where might the bird's cage have been? No one was willing to venture a guess.)

Eventually I came to the final room. Between the parlor and the little bedroom is an expansive *Study (?)*. Surely it was. Airy, light-filled, with space for a grand piano, violin, viola, and the traipsings of children, servants, students, musicians, dog. This was the perfect working room. And it was here that the ghost-bird on my shoulder finally fluffed his shining feathers.

In order to give Carmen the most time possible outside her enclosure, I typically bring her up to my writing studio while I work. This takes some preparation. I close the high windows, of course. Then I hide anything that I don't want to be nibbled, poked, stolen, thrashed, tromped on, or pooped upon. If it is winter, I turn off the radiator and cover it with a blanket so Carmen doesn't land on it and burn her delicate little feet. (Foot injury or infection is a common cause of death among captive birds, and foot health is something I watch in Carmen carefully, providing her with natural perches of varied width and roughness.) Once the room is ready, I place a fresh paper towel within reach (to swiftly wipe up pooplets), lock Delilah in Claire's room, then go down to open Carmen's aviary so she can fly to my shoulder and hitch a ride back up the stairs. Since reaching her wary adulthood, she has become suspicious of the stairwell, so I walk slowly and whisper calming sentiments as we climb. Once inside my studio, she is comfortable. This is her favorite room. So much to do! So many ways to pester me!

Carmen will set immediately to her computer work. She loves to ride on my hands while I type, but more than this, she loves to flap, jump, and scamper across the keyboard.

She tries to wedge her bill between the keys and the board, another example of the gaping behavior she uses to explore her world, and I worry that one day she will actually pop a key off or electrocute herself. It is difficult to believe that her computer play is all chance. She adds letters to documents, amends e-mails, erases e-mails, and *sends* e-mails before they are finished. She "likes" Facebook posts. Occasionally, her editing has improved a sentence, deleting one of the superfluous adverbs to which I am prone. Sometimes I stop to watch the movement of her rainbow-tinted feathers and her smart glistening eyes. After a minute she turns to look at me, as if to say, *Ahem! More typing, please! How am I supposed to spoil your pitiful attempt to work if you won't even do anything?*

Carmen oversees the writing of *Mozart's Starling.* *(Photograph by Tom Furtwangler)*

Once Carmen tires of the computer, she turns to her second favorite form of studio play: finding something that is dangerous and stealing it. Somehow she intuits immediately when I want to take away an item that she has made her plaything, and when I reach for it, she will fly tauntingly out of reach. I have learned that when she finds pushpins, thumbtacks, rubber bands, or other things that could kill her if she decides to swallow them, my best strategy is reverse psychology. I affect perfect nonchalance, then pretend interest in a *different* shiny little object, something harmless, perhaps a large, unswallowable paper clip. When Carmen flies over to investigate, I cover it up and tell her she can't have it. This will almost always inspire her to toss away the death-thumbtack and try to grab the paper clip. She'll peck at my fingers, wedge her bill between my knuckles to force them apart while squawking angrily, *C'mere, c'mere!* until she claims her prize (she always mimics when harried). If I pretend I want the paper clip back, she'll take it away and guard it.

To keep Carmen safely occupied, I have made a little toy box for her in my studio, as I did for Claire when she was a toddler. Carmen's is more of a toy bowl, really. It is filled with her favorite things: paper clips, binder clips, hair ties, sticky notes of various sizes and colors, buttons, a crinkly chocolate wrapper, a peanut in its shell, a teabag with string and tag. Every day I renew my vain hope that she will occupy herself with these things while I get some writing done. But the issue with Carmen is that she doesn't want to just play with these things, she wants to play with them *with me.*

They must sit on my desk, right at my elbow where she can choose one and "kill" it by shaking it back and forth, then thwacking it brutally against the desktop, a behavior she has exhibited since she was very young; she especially likes to attack scraps of paper in this way. I kept this in mind as I studied wild starlings outdoors and saw the behavior duplicated when birds captured insects, especially ones with large wings, like dragonflies or moths. The starlings would brandish their bugs and bash them against the ground until the wings were mostly shaken off, then they would eat the bodies of their now-wingless delicacies. (Carmen has never had the chance to catch anything with elaborate wings, like a moth or a butterfly, but if a little bug flies near her, I am shocked at how lightning fast she can catch it and swallow it down. She is, always, a wild bird.) Once Carmen kills all of her toys, she must show them to me one by one, hopping on my shoulder where she can drop a paper clip down my shirt or leave a sticky note in my hair.

After an hour or so comes my own favorite Carmen-time. Finding nothing more to play with, having spent long enough reading books over my shoulder and turning pages that I did not want turned, and having finished all of her e-mail correspondence, Carmen settles onto my shoulder, into the crook of my arm, or on my lap against my belly; she rounds her soft breast over her feet, fluffs and then unfluffs her feathers, and becomes perfectly still. Sometimes she will close her eyes; other times she will simply rest, entirely at peace. She might make a contented little sound, one I never hear from her aviary. It is a sigh-chirp, reserved for these moments of quiet snuggling. If I am still, I can feel her swift heartbeat. I will never tire of such moments. Comfort, rest, and unexpected consolation, shared so easily between two beings who grew from such seemingly different limbs of the taxonomic tree.

I envision moments such as these shared between Wolfgang and Star. I am certain that they were common. It is tempting to assume that in a respectable eighteenth-century home, the bird would sit decoratively in its rococo-style cage, interacting with the household humans only insofar as it took its flageolet lessons from the perch within. But pet birds of the time were allowed out of their cages and made part of daily life, much more so than most modern pet birds. While the typical pet-store birds today are not raised by hand, birds from eighteenth-century European bird sellers were generally stolen from their nests as young (alas, like Carmen) and raised in a home. They were accustomed to human presence, very tame, knew how to fly around a room without bashing into things, and on the whole made good family pets. Then, as now, the level of actual social interaction varied with the species of bird. Finches and canaries are tamable, diverting, and sing prettily but will always be a bit skittish. Species with higher plasticity of behavior—birds like parrots, starlings, and mynahs (the mimics, notably)—are capable of much more.

As far as we know, Star was the first bird Mozart brought into his married household, but he'd lived with many animals throughout his life. A field guide to Mozart family animals would include his childhood dog, Bimperl, several canaries, and the old horse he brought from Salzburg to Vienna whom he lovingly called the Kleper, or "nag." The Mozarts responded to their pets with a lively affection. When traveling in Europe, young Mozart would send newsy letters home that culminated with messages for the spaniel: "Cover Bimperl in a thousand million kisses, and pull her fluffy tail for me one hundred times, and fluff her fur until she barks!"

Star and Gauckerl, like other pets of the era, were at the center of a sea change regarding human attitudes toward nonhuman creatures. It remained a predominant belief throughout the eighteenth century that all organisms were created by God for human pleasure or edification. Birds sing for our delight. Cows dot the landscape to inspire feelings of serenity and, of course, to feed us. Even seemingly insidious creatures demonstrate Divine Wisdom and the love of God for Man: weeds exist to encourage human industry and the movement of our bodies for health; lice incite us to keep clean.

Before the eighteenth century, human relationships with animals were largely utilitarian—food, transport, work, and scientific research. While the aristocracy might have kept exotic pets, the bourgeoisie in general did not. With René Descartes at the helm of a fashionable rationalist philosophy, only humans were imbued with consciousness; animals were mindless "automatons," incapable of either thought or suffering. Their cries during scientific study or slaughter were likened by Descartes to the squeaking of a door on its hinges—nothing more. The philosophy of Descartes had become ingrained in the emerging sciences, particularly medicine, where it was used to justify horrific surgical research on live animals.

It was partly in response to Descartes that Rousseau published his *Social Contract* in 1762, just twelve years before Mozart brought Star home. Rousseau was born a Calvinist and earnestly converted to Catholicism as a young man, yet he came to believe that the highest insight into religious truth was not revealed but "natural"—found in the pageantry and harmony of nature. Animals became not just workers in the field of God, but windows into the character and workings of God. Rousseau wandered the countryside for peace of mind and heart, and there, amid the simple quietness of plants and the movements of birds, he experienced a "unity of all things" that anticipated the modern ecology movement. His work on animals is an outright rejection of Descartes. Animals may or may not be rational, but they are clearly sentient; they can suffer and so deserve to be treated kindly by humans. Rousseau laid the philosophical groundwork for a continuum between human and animal consciousness in the *Social Contract,* writing: "Every animal has ideas, since it has senses; it even combines those ideas in a certain degree; and it is only in degree that man differs, in this respect, from the brute."

It was the Rousseauian view of nature that largely inspired the culture of human-animal interaction in Mozart's time, when it became increasingly popular to keep pets in upper- and middle-class homes—dogs and birds in particular. (As mousers, cats were considered working animals and were most often kept by the lower and farming classes.) Pets were, more than ever before, brought into the house and the round of daily life, and it is this—direct daily contact with animals by ordinary people—that undermined Descartes more than any academic or lofty moral argument. People living with pets could see very clearly that these animals possessed consciousness worthy of human consideration. The pet-keeping of this period is often interpreted as being merely decorative, but Mozart and his starling were actually part of a great philosophical bridge leading from a disastrous lack of ethical standing for animals to the strong evolutionary case for animal consciousness in the nineteenth-century work of Charles Darwin and George Romanes, who in scientific language posited a continuity between humans and animals, both in body and in mind. Their work, along with Rousseau's *Social Contract,* led to a movement for the humane treatment of animals and eventually the original Humane Society.

Still, it would be a mistake to confuse this commonsense recognition of animal consciousness with a modern ecological view of wild animals in nature. Rousseau was an Enlightenment thinker but romantic to the core; for the public, the Rousseauian vision of nature's goodness, order, and harmony trickled down into an idealized, tamed, garden-like sensibility. Their pet-keeping reflected this.

Like the effort to make native birdsongs conform to a more human sense of proper musical structure, elaborate cages were patterned after the current taste in architecture. The finer cages were often crafted of mahogany, with silver or brass fittings and lots of tiny alcoves and turrets too small for a bird to enter. They were difficult to clean, decorated with toxic paint, and riddled with exposed nails—created more for ornament than for the health of birds. Tiny, portable cages that allowed singing birds to be carried from room to room were popular, an idyllic eighteenth-century version of the boom box. (I doubt the Mozarts had one of these, as there was plenty of music in the house already.)

Much of what we know about eighteenth-century bird-keeping centers on a study of portraiture. Professionally commissioned portraits from this period regularly show women and children with pet birds out of their cages. Birds are perched on the shoulders of their keepers, sitting on their fingers, playing with the ribbons on their costumes, eating bits of food from their hands or from the end of a proffered stick (like the Starbucks stirring sticks I used to raise Carmen). All these behaviors imply hours of daily freedom; birds that are primarily caged do not remain tame or tractable enough to interact with their keepers so gently and intimately. Any bird, even when it is raised by hand, will become wary of people when left alone in a cage. In Meredith West's modern study, the hand-raised birds that were kept on the porch rather than in the house lost their tameness among humans. But in these portraits, we see a teenage girl and a colorful finch perched on the same chair, looking at each other with ease and familiarity; a child and dove each holding one end of a red silk streamer; a woman in a parlor playing her tabletop piano for a chaffinch who stands gazing up from the toe of her silk slipper. (The woman herself looks aloof, bored, desperate for any kind of diversion, as so many intelligent upper-class women were.)

Elegant Lady with Miniature. An uncaged canary watches as the young lady dubiously assesses her lover's letter and miniature portrait. *(Michel Garnier, ca. 1790)*

For such paintings, all humans appear in their finery—little boys are scrubbed to shining, girls are skirted by froths of billowing pale silk. Men do not typically appear in portraits with birds unless it is a family portrait and the bird is among the children. Societally and artistically, the bird represents a sense of innocent pleasure; hence, it belongs in the realm of children and young women. The portraits convey the atmosphere of a pre-fall Eden, a proper garden of wonders for the young. Mozart, of course, was a man. But just because men are not seen engaging with birds in portraits does not mean that they did not do so in the privacy of their households. And Mozart, we know, was a bird of a different feather.

There is a reason that I have not been able to find a portrait that includes a starling from this period. Pet birds and family portraits were both symbols of social rank. Exotic parrots and finely trained canaries were expensive and indicative of a certain status. Starlings were common, native birds that any peasant could pluck from a nest. In a shop they cost just a few kreuzer and so were unfit for portraiture. It says a great deal that the status-conscious Mozart chose a starling for his pet—it means that he didn't want just *a* bird, he wanted *this* bird.

Where does all of this put Star in the Mozarthaus? For Mozart, there could be only one answer: Where the music was. Where Star could be watched, twitted to, flirted with.

Where he could join the music, pester the maestro, pluck at the violin strings, pull at the quill pen, tip the ink bottle. The Mozarts, always, made their own music, their own bohemian life. The elegy that Mozart composed for Star when he died shows such a keen understanding of the starling personality, such a personal knowing of his bird, that the two must have shared each other's company day to day. "He was not naughty, quite / But gay and bright." The little yellow statue of Star in the Mozarthaus should go in the study.

Did Star venture beyond the study? It is unlikely. Surely the servants had no time for such nonsense. In the parlor, Star might poop on the fine felt surface of the gaming table or upset the whist game by stealing a card. Guests in the parlor would not have appreciated starling poo in their freshly friseured tresses any more than my modern guests do in their hair. (If Carmen is out of her aviary when friends are around, we offer "poop shirts"—some old things of ours that Carmen can poop on—so no one's outfit gets mussed.) Besides, starlings get attached to a familiar place. Like Carmen with her aviary and its environs, Star was probably happy staying close to Mozart's cheerful study.

I can hardly breathe when I ponder Mozart's daily schedule and workload. His output was almost inconceivable. In 1784, the year he and Constanze moved to Camesina

House, he wrote six piano concertos, one piano quintet, one string quartet, two piano sonatas, two sets of variations for piano, and dozens of shorter compositions. In 1785 he composed three piano concertos, two string quartets, one piano quartet, a sonata and the Fantasia in C Minor for piano, the Maurerische Trauermusik (Masonic Funeral Music), operatic scenes for vocal trio and quartet, and several gorgeous parlor songs, and arranged the Mass in C Minor as part of a cantata.

We imagine him, like most writers, painters, and other artists, working in undisturbed peace and seclusion, with his dutiful spouse shushing the babes and guarding her genius husband's privacy. In the Mozart household, nothing could have been further from the truth. Constanze was a beloved partner to Wolfgang, but as Volkmar Braunbehrens writes in his brilliant biography *Mozart in Vienna: 1781–1791,* she was not "a good little housewife who spent the whole day doing laundry, cooking, and keeping the children away from her husband so he could have the peace to work. The Mozart house was a loud and restless one, but least of all because of the children, who played a minor role; it had more to do with music—pupils, house concerts, and rehearsals—and with Mozart's own need for constant commotion: conversation, laughter, visitors who were often houseguests."

We know that Mozart's house was full of music. He was running about Vienna, yes, but he was also at home for hours each day, composing, playing, teaching. It is said that when thinking he spoke aloud to himself in operatic recitative. Constanze had a splendid voice and sang as she knit lace, sewed, or chased and dandled the children. In the summer of 1788, Constanze's brother-in-law Joseph Lange (who had married Aloysia, Constanze's older sister and the object of Mozart's early romantic passion) brought a gaggle of theater folk, including two Danish actors, to visit the Mozarts. We have this wonderful diary entry from one of the Danes, a slice of Mozart home life:

> *There I had the happiest hour of music that has ever fallen to my lot. This small man and great master twice* extemporized *on a* pedal pianoforte, *so wonderfully! so wonderfully! that I quite lost myself. He intertwined the most difficult passages with the most lovely* themes.—*His wife cut quill-pens for the copyist, a pupil composed, a little boy aged four walked about in the garden and sang recitatives—in short, everything that surrounded this splendid man was* musical!

The diarist was an enamored guest, and Wolfgang surely took a moment to show off, but nearly all reports of daily life at the Mozart compound are similar: musical and domestic chaos, and the composer happy within it. There were the children, upon whom Wolfgang doted, and the dog, and the students. Mozart taught female students in their own homes, where they were chaperoned, as was customary, but male students came to his studio, and particularly gifted young men would stay on for months, perhaps a year, and live as part of the household, treated as family. Mozart not only managed to compose amid all this chaos; by all reports, he *preferred* it. The background noise and bustle was something to work against. It is said that when Wolfgang was composing music in his mind, his outward actions changed little, but something about his countenance became for a moment a touch distant, as if he were listening to a faraway birdsong, before his quill sped along again. Star would not have been a distraction at all—at least, not an unwelcome one, but rather a strand in the splendid, essential, almost divine chaos.

Leopold made just one extraordinary visit to Wolfgang and Constanze after they were married. He arrived in March of 1785 for a ten-week stay, prepared to be appalled at the state of the household and marriage. Instead, he was impressed with Constanze's economy and enthralled with the liveliness of the home. He wrote to Nannerl, whom he had left alone in Salzburg: "It is impossible to describe the rush and bustle. Since my arrival your brother's piano has been taken at least a dozen times to the theater or to some other house." This happened constantly—the fussy maestro preferred his own instrument for performances, and so it was moved every two or three days.

Earlier that year, Mozart had composed a set of quartets

for his friend and mentor Joseph Haydn and was anxious for the maestro's opinion before he ventured to formally dedicate them. During Leopold Mozart's stay in the Domgasse apartment, the quartets were performed at a small party for Haydn, with Wolfgang playing viola, and Leopold Mozart himself playing first violin. After, Haydn famously commented to Leopold: "I swear to you before God and as an honest man that your son is the greatest composer I have ever known in person or by name" (and in a jarring departure from the rest of the Mozarthaus museum's simplicity, these words are painted in slender period calligraphic style on the salon wall). In my imagination, Leopold's narrow face is almost permanently lined with an anxious semi-scowl, his dark version of the Mona Lisa's smile. Still, the pleasure he felt at this compliment could not have been hidden.

The night was a glorious one for Mozart. Haydn loved the quartets, his father was proud, the parlor was filled with glowing candles and the swooshing whisper of brocade. During my visit I walked to the windows of this open sunny room, which overlooked a narrow, cobbled pedestrian alley, sweet buildings, and my favorite little café away from the busy-ness of the Graben, surrounded by pink autumn petunias. I closed my eyes and heard the music.

If my suspicions are correct, there was another layer to the musical offerings that night, beyond the usual quartet instrumentation. Mozart's study and, if I am right, Star's cage were just one room away. An acclaimed neuropsychologist wrote that "only humans have a natural, or innate, inclination to engage with music." I read this aloud one evening at the dinner table and everyone cracked up. Clearly this scientist has never lived with a starling. When live music is played in our house, Carmen is constitutionally incapable of silence. She jumps and flutters and then settles into singing with full starling exuberance. Whistled cadenzas, warbles, *Hi, Carmen! Hi, honey!* Wine-stopper sounds, more whistles. Wildness, joyfulness.

There is plenty of opportunity. Claire is a gifted musician, playing for hours a day on her cello but also on piano, mandolin, and guitar. I play the Celtic harp and a little violin and piano. Tom—well, Tom plays the cowbell. Usually, we just revel in Carmen's participation, hearing it as an element of the music—par for the course in our slightly eccentric household of writers and players. But it can be a nuisance when Claire is trying to make a cello recording for an audition or give a serious home performance. Sometimes I'll take Carmen up to my study when Claire needs to make a clean recording, but once we attached a note to a recorded audition for a prestigious festival. Claire had played perfectly and nearly cried when she heard the overlay of *Hi, Carmen! Kiss me!* on top of her Tchaikovsky. *Please excuse the bird sounds. We have a pet starling. Mozart had one too.*

It's not just Carmen. This is what starlings do—*they join the music.* What happened the night that Haydn's quartets were performed? Did the Mozarts toss a blanket over Star's cage to discourage the loud whistles and warbles that the music surely incited? Did they close the door to dampen Star's song? Certainly not. Leopold loved birds at least as much as his son did. I would bet a hundred gulden that they all laughed, opened wide the doors between the rooms, and enjoyed another strain of the household orchestra.

Six

HOW THE STARLING KNEW

Though I did not attempt to teach Carmen particular words or phrases while raising her, I *did* try to teach her the motif from Mozart's Piano Concerto No. 17 in G, the musical phrase that, we know from Mozart's notebook, Star could sing. I wondered if it might be a tune that starlings in general responded to, and in any case, it would be a wonderful element of the narrative to say that Carmen learned Star's tune. I imagined the YouTube video that we would produce: Carmen singing alongside images of the maestro, murmurations of starlings flying in the background, and eventually a full orchestra picking up the theme. So even when Carmen was a baby bird tucked into her cottage-cheese tub-nest on my desk, I played the motif from Mozart's concerto for her on my violin at least thirty times a day; I hummed the tune while tidying my study; I looped an orchestral recording on my iPod and let it run while I was at the grocery store. Carmen was raised on a steady diet of homemade starling mash and Mozart.

As it turned out, Carmen had no intention of learning the motif. But by the time she was two months old, she did take a keen interest in learning the violin. She is the only creature on earth who has ever seemed to take pleasure in my playing. (Once I was practicing with the window open, and *three* neighborhood dogs began to howl.) Her favorite place to study violin was from her perch at the tip of my bow. At two months, a full-size bird, she weighed just two ounces. Still, it is difficult to play with a starling on your bow. Now, as an adult, Carmen prefers to perch on the scroll of the violin and gape between the strings; she places her bill between two strings, opens her mandibles wide, and then pulls her bill out, so both strings ring. This seems to delight her. But she will not learn the Mozart motif. Starlings are among the few songbirds who continue to learn new vocalizations year after year, so I am not without hope that she will surprise me one day. It is a sweet phrase, and Star could not have chosen a lovelier tune to share with his new owner.

I attended the Seattle Symphony's performance of the Concerto in G recently, with internationally acclaimed virtuoso Imogen Cooper at the piano and conducting the orchestra as she played. She walked through the big doors of the performance hall and onto the stage like an oak draped in red sateen—statuesque, strong, rooted. As this concerto begins, we in the audience scarcely have time to adjust to our surroundings before Mozart tosses us headlong with his music into the full current of human emotional possibility, yet he manages the swift transitions with such *beauty* that we do not think to resist. Mozart believed, always, in beauty and in harmony and would not sacrifice either, no matter how dark his themes. He wrote this out in a famous 1781 letter to Leopold that is now taken as an articulation of his musical philosophy and a foundational statement of Viennese classicism: "Passions, violent or not, must never be expressed to the point of disgust, and Music must never offend the ear, even in the most horrendous situations, but must always be pleasing, in other words, always remain Music."

The concerto has three movements, the allegro, the andante, and the allegretto. In the first, the allegro, Mozart progresses from one musical idea to the next without restraint and with practiced effortlessness. The effect is a nuanced joy that ventures into unexpected keys with such fluency that we almost forget to be unsettled. The woodwinds carry much of the discourse here; we modern audiences might not even notice it, but in Mozart's time a strong woodwind voice was unusual. After all the sweet activity of the allegro, the dark serenity of the andante falls over the audience like a shroud. The opening is an ethereal string theme, which after about twenty seconds abruptly stops. Just stops. The oboe and bassoon take up the silence and sing behind a floating solo flute. Finally, the piano enters, entirely alone. The dramatic pauses continue, reminding us constantly that we are in the hands of a master operatic composer—operatic drama will emerge increasingly from here on in Mozart's concertos and symphonies. Now it is the piano that sings this concerto's wordless aria. There are harmonic surprises and further forays into unusual keys. I cannot close my eyes, because I don't want to miss the oak-trunk movements of Imogen Cooper at her instrument, conducting with limited, but dramatic, movements of her body, arm-branches, and eyebrows. But even with my eyes open, I feel I am surrounded by forest imagery—earth, mist, a veiled, enchanted-but-dark place. And the flute, to me, is Pan's.

There is no rest. The allegretto leaps immediately into the relief of G major and the first notes of the starling's motif. The shadows disperse. Instead of the expected rondo, Mozart dispatches five variations on the theme and then, in the finale, runs away with it in a prodigal fantasia in which Star's motif surfaces over and over against the riverine flow of the piano cadenzas.

They say that birds prefer Mozart above other composers, and perhaps this is true. But not Carmen. She prefers Bach and bluegrass. Based on the exuberance of her reactions, she even has a favorite band—Greensky Bluegrass. When this beautiful concerto of Mozart's is playing, she will sit impassively on my shoulder, almost yawning. But when the final movement begins, she is excited. She jumps down to my hand where she can look me in the eye. *Hi, Carmen! Hi, honey!* she calls before breaking into her own shrill starling aria. There is something here for her. There is something here for everyone. In the program notes to the Seattle performance, the commentator suggests that the andante movement's "nod toward melancholy is swept away by the next animated variation." But in my hearing, this is not true. The melancholia remains, perhaps more romantic than existential, but lingering within our listening experience. The exuberance of the finale contains the darkness, enfolds it, shelters it, redeems it. But does not forget it.

Like most of Mozart's music, this concerto is written on two levels—one for the simple enjoyment of the musically uneducated ear, and one for the musical adept. Mozart could not have maintained his public popularity and sustained his own genius-level interest in any other way. In a famous anecdote, Emperor Joseph was present for the first performance of the opera *The Abduction from the Seraglio* and remarked afterward to the composer, "Too beautiful for our ears and an extraordinary number of notes, dear Mozart," who was said to have replied, "Just as many, Your Majesty, as necessary." The exchange as imagined by Peter Shaffer for the play and movie *Amadeus* is more caustic and has become something of an anthem to the creative spirit:

> JOSEPH: There you are. It's clear. It's German. It's quality work. There are simply too many notes. Do you see?
>
> MOZART: There are just as many notes, Majesty, neither more nor less, as are required.

I agree with many scholars' skepticism over the authenticity of the exchange, but it does capture an honest conflict for Mozart, a young composer whose head was nearly exploding with new musical ideas, a man who required imperial support but whose audiences, both public and sovereign, were not quite ready for the full force of his originality. Biographer Maynard Solomon notes that while the story may not be reliable in its fine points, it is nevertheless true that in the next round of imperial sponsorship for German-language opera, Mozart was not employed to write anything except a minor comedic singspiel.

By now, the fact of the starling's ability to mimic a simple musical phrase should be no surprise. But rumors over *how* the bird learned Mozart's refrain have run wild. In short commentaries and recording liners and program notes for this concerto, I have read emphatic but unsupported statements that the starling taught the motif to Mozart. That Mozart taught it to the starling. That Mozart heard the starling whistle a folk song that sounded like his motif. And, most astonishing, that "Mozart trained a pet Starling to whistle, in its entirety, all of one movement of his G major piano concerto, though the bird consistently sang two notes a bit flat." Not only is the "all of one movement" claim a shocking hyperbole—even for a starling—but we know from the notations in Mozart's book that Star sang one note sharp, not two notes flat. What respectable starling would be caught dead singing *flat?*

The whimsical suggestion that Star taught the motif to Mozart—that he sang a song of his own adapted by Mozart for the concerto—appears in print with some frequency. Such claims are proffered by folk who are speculating with just a tidbit of information. Though Mozart will, as we'll see, incorporate starling-esque cadences and personality traits inspired by Star into later work, this concerto was completed more than a month before Mozart brought his bird home, and so for this composition such a proposal remains nearly impossible. There is, however, a sweet children's book that fictionalizes the theory delightfully. In *Mozart Finds a Melody,* by Stephen Costanza, Mozart is facing a deadline and is plagued by a bad case of writer's block:

> For the first time in Wolfgang's life, the famous composer was at a loss for a tune. He tried every trick to get his imagination going. He sang standing on his head. He played his violin in the bathtub. He even threw darts at the blank music paper. Alas, nothing worked.

The bird sings notes that start to coalesce into a theme that Mozart can use, but just as Mozart is ready to put quill to paper, the bird flies out the window. The distraught composer must find the bird in order to finish his tune. The role of starling as muse is authentic, but the idea that the bird taught *this* tune to Mozart is far-fetched.

But what *is* true? How and when did Star learn Mozart's motif? There are really only two overarching possibilities: either Star learned the motif *before* Wolfgang purchased him at the shop, as in my introductory tale, or he learned the song *after* the purchase, while living with the Mozart family. There are problems with both. If Star could sing the motif before (or at the time) that Mozart bought him, then how had he learned it? If Star learned the motif later, then how could Mozart have recorded the tune in his notebook at the time he bought the bird?

The answer to this question was a kind of holy grail to me while I traveled in Vienna and Salzburg. I believed that by the time I finished my researches there, examined documents, wandered Mozart's homes, talked with experts, and let daydreams trail through my brain under the influence of the Austrian landscape, I would, somehow, have uncovered the solution to this lovely musico-ornithological mystery. It didn't *matter,* I knew, in the scheme of things. Why, really, when I know the essentials of the story, should I care about the arcana?

I can answer only in the way thousands of seekers over hundreds of years have answered their own version of such a question: I care with the brightened curiosity of one who loves a subject for no rational reason, but who loves it nonetheless, and prodigally. This is the ardor of the academic Austenologist who believes that if she looks beneath the floorboards of the right dusty attic, she will find the diary entry explaining why Jane Austen rejected her one marriage proposal the day after she'd accepted it; of the birder in Costa Rica tiptoeing through trails of biting ants and fer-de-lance serpents in hopes of glimpsing a rare hummingbird that no one has seen for fifteen years. I could list such loves forever, the sort that visit our imaginations on the cusp of the impossible but that we cannot erase from our minds. We follow the trail with whatever bread crumbs we can gather, with hope, with love, with an almost magical combination of urgency and patience. There were just enough crumbs in the Mozart story that I felt confident that, with enough sleuthing, the details of just how the Mozart-and-Star story unfolded would fill my grail chalice. Naturally, that is not at all what happened.

As far as objective facts go, in all of my Austrian snooping, I uncovered little more than the broad strokes I already knew. Mozart's own catalog of work tells us that the concerto was completed on April 12, 1784. It has long been believed by musical historians that the piece was meant to be a strict secret, not performed publicly until mid-June of that year, when Mozart conducted it with a small chamber ensemble that included Barbara Ployer—his gifted young student for whom the concerto was composed—at the piano. This recital took place before a small, elite audience at the von Ployers' country residence, where the great doors might have been opened to the garden of chestnut trees, allowing the cool evening breezes to lull the attendees into an even deeper romantic state than the music would normally induce.

We know from Mozart's expense notebook that Star was purchased on May 27 and that he could sing the line from the Piano Concerto in G. It was Mozart's habit for a time to record all his expenditures in this small booklet. The purchase just prior to the starling reads, *Two lilies of the valley, 1 kreuzer,* making Mozart, with his acquisitions of flowers and birds, appear to be even more romantic than he actually was. Previous and future purchases are more prosaic and include staff paper, ink, books. The notebook was used as a primary reference in an early 1798 biography of Mozart by Franz Niemetschek, for which Constanze provided source material. More thorough references to the notebook, with facsimiles of some pages, including the ones with the starling references, appear in a thematic catalog of Mozart's ephemera gathered in 1828. In the commentary for this facsimile, it is mentioned that the musical notation recording the starling's song was written *zugleich folgende,* or "immediately following," the record of the bird's purchase, implying that he recorded the starling singing at the same time that he bought it.

If not for the notebook, it would be easiest to argue that Star learned the motif once he was home with the Mozarts. This is the way of starlings. They involve themselves in the daily sounds of their flock—whether that flock is made up of birds, humans, or violins. Like any other starling worth its feathers, Star would simply have absorbed and mimicked favored sounds from his setting—and what a setting he had! In this scenario, Star simply picked up the tune as it was practiced or whistled by Wolfgang in his study. And what bird (besides my contrary Carmen) wouldn't love to sing the merry rondo of Mozart's seventeenth concerto? It is a sweet, self-contained, chirpy little refrain—perfect for a starling.

Even so, Meredith West argues in her seminal 1990 paper that Star could already sing the motif when Mozart bought him in the shop. She relies heavily on the evidence provided in what we know of Mozart's little expense book combined with her research on the nature of starlings:

> Given our observation that whistled tunes are altered and incorporated into mixed themes, we assume that the melody was new to the bird because it was so close a copy of the original. Thus, we entertain the possibility that Mozart, like other animal lovers, had already visited the shop and interacted with the starling before 27 May. Mozart was known to hum and whistle a good deal. Why should he refrain in the presence of a bird that seems to elicit such behavior so easily?

When I spoke to Dr. West recently, nearly thirty years after the publication of her paper, she told me that she still feels this is the most likely way for events to have unfolded. It is entirely plausible; Mozart was drawn into a shop full of singing birds. How could he not be? Wolfgang had kept birds during his childhood and loved them. The exuberance of the starling's song he now heard—so different from the canaries he had grown up with—piqued his native curiosity. In her paper, West supports her theory that Mozart was attracted to this particular bird for more than its song, noting that, like a starling, Mozart himself was attuned to nonverbal communication and to visual cues. She invokes the feeling that overwhelmed Mozart at a performance of *The Magic*

Flute, where he found the "silent approval," as he wrote to Constanze—his *feeling* that the audience members were attuned, involved, in love with his music—even more gratifying than their enthusiastic applause. Mozart observed in his audience the same attitude assumed by a starling when something captures its interest. When you talk to a tame starling, it jumps as close to you as it can get, tilts its head, and listens. So gratifying! Mozart was well aware of his own gifts, yet craved attention and approval for his music. When visited by Mozart in the shop, Star gave the composer these very things. As West writes:

> To be whistled to by Mozart! Surely the bird would have adopted its listening posture, thereby rewarding the potential buyer with "silent applause."

While the canaries and finches were flying around their little shop cages, thrashing and singing within their own little worlds, here was an iridescent bird looking right at Mozart and seeming to wonder, *What will this man say next? What will he sing? What will he whistle?* Wolfgang could not resist such an agreeable audience. He whistled his dear little phrase. Whistled it again—four times, five. Carmen takes much longer than this to learn a new word, but a particularly gifted and willing starling mimic could begin to pick up the phrase that fast. Perhaps Mozart visited a couple more times, unable to resist the call of this flirtatious new friend. And when Star had learned a passable version of the motif? Well, how could Wolfgang not take such a friend home?

My own imagined tale, recounted at the beginning of this book, is a variation on West's theme. Here Star has learned the motif before purchase, but Mozart did not teach the bird his tune. Instead, he was drawn into the shop by the sound of the bird singing it, as if following the fragrance of a freshly baked Viennese plum cake. This, too, is possible. The bird catcher's stall was almost certainly on the Graben, a vital commercial district bustling with shoppers and musicians day and night. Mozart composed the concerto when living at 29 Graben, the apartment's windows just above the string of markets and sellers of all kinds. He was still copying the music over in May, when the windows would have been open to the spring air after a dust-settling rain and the musical strains might have filtered out into the world and perhaps into a starling's little ear. There is no reason that Star could not have picked up a bit of the motif and played it over on his starling tongue until it came out nearly in the form that Mozart had composed. And once it did, there would have been no stopping the repetition—starlings *love* to share freshly learned mimicked sounds.

But scholarship that has surfaced since the publication of West's paper suggests another, more plausible way by which Star could have gleaned the tune before Mozart met him. We now suspect that the concerto might have had a somewhat earlier public debut at a prestigious concert with Emperor Joseph II in attendance at the beautiful Kärntnertor Theater (now the footprint of the famous Hotel Sacher) on April 29. This would be in line with Mozart's typical eagerness to place a piece before the public eye, even if the ink on the pages wasn't quite dry. It would not even have been the newest work by Mozart performed that night, for the evening's program was ordered in part to feature the virtuoso violinist Regina Strinasacchi. Wolfgang wrote to his father on April 24, a Saturday: "We have the famous Strinasacchi from Mantua here right now; she is a very good violinist, has excellent taste and a lot of feeling in her playing." Mozart loved to compose specifically for particular virtuosos or voices—some of his pieces were so difficult that they could be performed by only one exceptional artist. He continued astonishingly: "I'm composing a Sonata for her at this moment that we'll be performing together Thursday in her concert at the Theater." It takes a lot of temerity to compose a piece just days before it will be performed before the emperor. It has become a famous (and true) Mozart anecdote that he played the piano part this night from an unfinished penciled score, improvising the cadenzas and conducting at the same time.

It is now cautiously believed by many scholars that the Piano Concerto in G debuted publicly this night as well, also with Mozart at the piano. It would certainly have given the tune a chance to get out on the street before Mozart met and bought Star. The theater was just a ten-minute or so walk from the district where the bird catcher's shop likely stood, down a cobbled street past the Hofburg Palace at the end of the Graben (near the current opera house).

And the vector from performance hall to starling? The

famous whistlers of Vienna, of course—the everyday street-walking people who soaked in the surrounding music and then whistled it back out again. Humming and whistling filled the streets of eighteenth-century Vienna. There were no radios, no iPods. There were only these public performances and the repetition of a strong theme, as in this concerto where the catchy motif that occurs repeatedly within the allegretto and is reprised in the finale was an intentional gift from the composer to the audience; it gave people something memorable to take home.

It is said that from childhood, Mozart could memorize whole symphonies after one hearing well enough to write them down note-perfect. (In 1770, Leopold was traveling with his fourteen-year-old son and wrote home to Anna Maria after attending a performance at the Sistine Chapel, "You have often heard of the great *Miserere,* in Rome. It is so greatly prized that the performers in the Sistine chapel are forbidden on pain of excommunication to copy it." Leopold was not exaggerating. He went on. "But we have it already. Wolfgang has written it down.") Most people, of course, did not have this skill. Composers—and Mozart was a master at this—would create a refrain not just for the art of their music, but for the heart and ear of their public, who would have an easy-to-remember phrase to carry with them and hum in days to come. Star could well have heard Mozart's tune whistled from the street by a concertgoer. Even more likely, since starlings learn best from personal interaction, someone who had attended the concerto performance might have stopped into the shop and whistled the phrase to the bird, who then learned it quickly.

All of these possible pathways to Star learning the motif involve educated imagination but no exaggeration of known facts, Mozart's biography, the musical timeline, or starling capabilities. All are within the realm of the intelligently possible. Yet there is a gray area here. What if Star learned the motif after Mozart brought him home, and Mozart simply added the musical notation to his notebook later? That would explain everything.

But here's the catch: If the bird learned the song after he came to live in the Mozart home, Wolfgang would have had to record the purchase of the bird, then go back to his notebook and record the mimicked motif later, once Star had started to sing it. This doesn't seem like such a hard thing, except for the fact that Mozart was a dismal record keeper. He did maintain a good catalog of his finished music, but other than that he didn't keep track of much at all. He had no diary, just the listing of purchases in this little book, and within a year even this effort would lapse—the last pages of the expense notebook were given over to practicing written English.

Like many on Mozart's trail, I long to get my hands on that booklet. Was there a change in ink used? Odd spacing? Anything that might indicate that the musical notation was made at a different time from the purchase notation? Sadly, we do not have the actual expense book to search for further clues—its whereabouts, if it still exists, are unknown. It may be hiding in the labyrinthine bowels of some German university or falling to dust in the basement of one of Constanze's distant descendants. For practical and scholarly purposes, the notebook is simply gone.

But recall that in the facsimile pages copied directly from the notebook, the commentator clearly stated that the musical entry appeared *zugleich folgende,* or "immediately following," the recording of the expense for the bird, which would support the idea that the notation was made at the time Star was purchased. Thus, what we can know of the ephemera does seem to point to the mystery of a bird in a shop having learned Mozart's song before coming to live in the family home. In truth, though, there is no way to know for certain, and I have come to accept that those who claim otherwise are overreaching.

When I first drafted this chapter, I included my best guess as to when and how Star learned Mozart's tune. But I deleted it. I didn't even put it into my Cutting-Room Floor File for possible revisiting. I deleted it into the ether, where it belongs. My grail chalice has been filled with an elixir that is perhaps headier than the wine of fact—it is filled with swirling, essential uncertainty and the difficult, mature task of dwelling in such a state.

And yet, I believe inquiry into what we do know to be true matters tremendously. History like this that has a basis in fact yet lies outside of memory is necessarily subject to imagination. To make it comprehensible, to make it real to our artful human minds, we tell it as a story. But when we force our will, rather than our intelligence and honesty, on a historical story, it loses its reality and is diminished. At its heart, the story of Mozart and Star is beautiful, meaningful, and true. It is belittled and made false by exaggeration and mere rumor—the starling did not teach the motif to Mozart; the starling did *not* sing an entire concerto. But we can perceive the sweet and authentic center. We know that an unusual friendship between one of history's most loved composers and one of the world's least loved birds began in May of 1784, when, in one order or another, Mozart wrote the starling's song in his book, conferred his judgment—*Das war schön!*—plunked down his kreuzer, and took the bird home, smiling like a jackass eating thistles.

Seven

CHOMSKY'S STARLING

C'mere, honey! Tom, Claire, and I were standing in the kitchen as Carmen called from her aviary. We all stared at one another, silently. As usual, Claire voiced the brave thought first: "She made a sentence." No one spoke after that except Carmen, who said, *C'mere, honey!*

In the past, Carmen had mimicked *Hi, honey,* and *C'mere,* but never *C'mere, honey!* So, yes, it seemed that she had just combined words and phrases, albeit simple ones, into a new pattern that made sense. Instead of mimicking a sentence she'd heard, it appeared that she had done something of a different order: she'd made a sentence of her own. Finally, I ventured uncertainly, "That wasn't a sentence," at least not an intentional one. Either she'd put the words together in a new way that happened to make sense (she jumbled up her repertoire all the time), or we had unwittingly been saying, "C'mere, honey," to her, and she'd latched onto it. Is it *impossible* that she meant to create a small sentence? Probably, yes. But given what I've learned, both from Carmen and from the current science on birds and language, I will never underestimate the possibilities —for starlings, or for any of the beautiful, bewildering voices in our more-than-human world.*

It is easy enough to find a starling voice to contemplate. At any time of year, we can meander down the sidewalk and stand beneath a starling settled on an electric wire—its favorite urban perch. It is tempting to conclude that the long jumble of whistles, gurgles, and clicks we hear is nothing more than a haphazard mess of sound. But it isn't. With a bit of attention, it is not difficult to tease out the sequence. Starlings *will* sometimes sit there and rehearse seemingly random whistles and imitations and various kinds of chatter. But in a full-fledged song bout, where the bird throws its head skyward and lets loose for five seconds, or twenty, or up to forty-five, the song is divided into easily recognizable sections that follow a predictable pattern. The first is the whistle. This is a series of long, wild, teakettle-ish sounds, ranging stormily up and down the tonal spectrum. (It leads

* The evocative phrase "more-than-human world" was first used twenty years ago by philosopher David Abram in his book *The Spell of the Sensuous.* It has since become a well-loved expression in the ecological literature.

to some confusion among those who encounter captive starlings like Carmen and are waiting for them to mimic something; "She's imitating the teakettle!" visitors delightedly announce when they hear Carmen's whistle—alas, she imitates a lot of household sounds, but this one is just a good old wild-starling voice.) After the whistle there is a quick pause, then the bird will break into its own personal repertoire of sound phrases, some of them mimicked, some of them learned from other starlings, some of them invented in its singular starling brain, all of them gathered into a sequence that is entirely unique to this individual bird. The literature claims that a single bird will have somewhere between six and thirty individual sounds in its repertoire, but based on Carmen's performance, I suspect the number of phrases for a gifted wild bird is beyond our current expectation. To me, Carmen is a wondrous mimic, but in the starling world she is simply average. She has fifteen phrases in her mimicked repertoire (that I recognize—there are probably more), and she is learning more all the time. Though females are fine mimics, a gifted male will mimic more phrases—and of higher complexity—than the average female. I have a hard time believing that among the millions of starlings in the world, thirty is the upward limit for a precocious male singer.

After the initial whistle and repertoire sections of the song bout, there is another swift pause. A male bird will now perform a series of clicks and rattles. This is one vocalization that females do not usually employ, and it's a fair way to tell if you are listening to a male or female bird.* The song will end with a whistled crescendo, different from the initial section in that it is quicker, more direct, less varied, and usually louder. That is the full starling song: whistle, repertoire, rattle, crescendo. It may differ a bit here and there, but it is dependable on the whole. This is no sweet-fluted wood thrush strain. This song is crazy-wild. It's gorgeous, in a loopy starling way. And it springs from a little bird with no idea that it has, in the pattern of its song, dropped its shining little body and brain into a turbulent academic debate—one that is both scientific and poetic. Here biology, language, art, music, consciousness, and—yes—human ego mingle, dance, and clash.

In the 1950s, a brilliant young linguist at MIT named Noam Chomsky was doing some hard thinking about the nature and uniqueness of human language. When Chomsky set to work on the topic, all of the social sciences, including the academic discipline of linguistics, were dominated by the tenets of behaviorism, which held that the only proper arena of psychological study for both humans and animals was observable, measurable, external behavior. Behaviorism had gained ascendance within the sciences in the previous

* This is *generally* true. It seems that about one in ten female starlings does have the ability to make these sounds. Or perhaps all female starlings have the physiological ability and only this smaller fraction makes use of it.

decade due to its ability to quantify human and animal responses in experimental settings, making so-called soft sciences, like psychology, hard, like mathematics and chemistry. Work on something as murky and unknowable as consciousness was relegated to a scientific backwater while the behaviorist model garnered grants and publication in the best journals. On the language front, behaviorism's progenitor B. F. Skinner decreed that children learned words and grammar by being positively reinforced for correct usage. The payoff most often came in the form of a response—eye contact, attention from an adult, or the chocolate cupcake the child had requested using the proper words. Such rewards, for Skinner, were akin to a lab rat getting its Purina Lab Chow pellet after pushing the right button.*

Chomsky penned a scathing review of Skinner's *Verbal Behavior,* and like Kant responding to Hume, woke from his

---

* Purina Lab Chow, now called LabDiet, is a real thing, as I learned in college. One day I was trying to find the office of a psychology professor with whom I had a meeting and walked past a room with a sign on the door: ANIMAL LAB. I didn't even know there was an animal lab at my tiny liberal arts school. Inside, I found drawers full of rodents, and I opened every one; in the last drawer, there were six fluffy Siberian teddy-bear hamsters. This species of hamster does not do well in groups, especially in small spaces; they become aggressive and even cannibalistic. The hostility of this group was focused on one sweet, cream-colored, long-haired male, who was covered with bites, had part of one ear nibbled off, and was bleeding from several wounds. I did not think, in the moment, about the fact that I might be ruining someone's thesis research. I just plucked up the bleeding hamster and put it in the left pocket of my jacket. I looked quickly around the room. Nothing; no one. So into the right pocket I dropped a handful of the hard, square kibbles from the giant bag of Purina Lab Chow that slumped against the wall. Diogenes the hamster (I was a philosophy major) lived many happy years.

"dogmatic slumber" to promulgate his own view. He pointed out that children create grammatical sentences following patterns that they have never heard, proving that instruction and reward are insufficient to explain the complexity of human language learning. He came to believe that humans are endowed with a faculty for language, a "language organ," as he misleadingly described it (since it cannot be said to be housed in a discrete physical structure), that contained a universal and immutable set of rules shared by all human languages, no matter how varied these languages appeared on the surface. This is Chomsky's Universal Grammar, or UG. We generate meaning by using these linguistic rules to combine words, build clauses, and incorporate these clauses into longer sentences.

More than this, humans don't make sentences solely by adding to them bit by bit, as if a sentence were a rat growing a longer tail. Instead, we often embed phrases within sentences, a process linguists call recursion. Take the discrete phrase, also a short sentence, *Mozart played the violin.* We could add to it in linear fashion: *Mozart played the violin and liked birds.* Or we could employ the syntactical device of recursion to embed the added information into the original sentence: *Mozart, who liked birds, played the violin.* And we could go on: *Mozart, who liked birds, and in fact composed with a starling perched on his shoulder, played the violin.* Perhaps: *Mozart, who liked birds, and in fact composed with a starling, which had shockingly iridescent feathers, perched on his shoulder, played the violin.* We could do this forever, limited only by breath and the capacity of memory. Many animals communicate in the linear fashion, we know. Cetaceans, elephants, winter wrens—all add complexity to their vocalizations by taking a starting motif and adding to it, then adding some more. But using recursion, humans are able to make meaning through sentences that are more like blossoming peonies, growing from within. Chomsky came to believe that the capacity for recursion is not only unique to humans but *the* defining characteristic that sets human language apart from all other forms of communication among living beings.

Aiming to bolster his theory, Chomsky coauthored a 2002 paper published in *Science* with Harvard biologist and psychologist Marc Hauser and Tecumseh Fitch, a linguist at the University of St. Andrews in Scotland. In "The Faculty of Language: What Is It, Who Has It, and How Did It Evolve?," the three researchers sought to identify the distinguishing features of human language and its development. They scoured the literature to compare aspects of human and animal communication, eliminating shared characteristics one by one. Memory, for example, is critical to language, because, as Hauser put it in an interview, "If you couldn't keep in mind several pieces of a sentence, you couldn't understand anything. But memory's not specific to [human] language." Some animals use memory in communication too. So memory was off the Unique to Human Language list. The researchers continued down their catalog of linguistic features until they were left with just one property that could be said to define human language alone: recursion.

To test their theory, Hauser and Fitch created a study to see whether a nonhuman primate with complex vocal capacity could recognize recursive patterns. They selected the cotton-top tamarin, a tiny, beautiful New World monkey originating in Colombia. They chose the species for its vocal complexity, but if you have seen a cotton-top tamarin, you might wonder if it wasn't chosen for its cuteness. They are among the earth's smallest primates, with dark faces surrounded by long manes of white, feathery fur. Their gaze is intense, open, curious. Somehow, the look on their faces seems to say, *Sure, let's study recursion! Have at it! What do we do first?*

Hauser and Fitch started by creating a simple artificial language. The words were made up of short sounds pronounced by either men or women, sounds like *li* and *mo*. The difference in pitch between the male and female voices made it easy for anyone listening to tell the two groups apart. Then the scientists combined the sounds into patterns—short sentences, really—using two rules. The first rule was very simple: a female sound would always be followed by a male sound. With A and B representing the two sound groups, this simple rule would yield short sentences such as ABAB. The second, more complex rule embedded a female-male sound pair within one of the AB pairs, with a resulting A(AB)B.

To create a baseline from which to judge the tamarins, the researchers first tested humans, innocent young Harvard undergrads, in fact, to see whether they could learn to recognize both patterns. After listening to thirty sentences, the students were tested on new ones. More than 80 percent could correctly recognize the recursive sentence patterns. Then Hauser tested the monkeys in his lab. The tamarins listened to recordings of the patterns over and over in the evening, and in the morning, the recordings were played for the monkeys again, along with the recordings that violated the patterns. When the monkeys recognized a novelty, they would turn toward the speaker and listen intently. According to Hauser, the monkeys noticed when the ABAB pattern was violated but were not able to pick out violations to the A(AB)B pattern. Chomsky was not the least bit surprised.

In a disconcerting footnote, several years after the publication of the tamarin study, Hauser's professional colleagues began to question his conclusions—and his research ethics. Hauser was a well-known tenured professor at Harvard in 2007 when the school began an internal investigation around possible scientific misconduct. By 2010, the allegations came to public light, and scandal erupted around him. While Hauser had honestly reported that the tamarins could not recognize recursive patterns, the U.S. Department of Health and Human Service's Office of Research Integrity found him guilty of several counts of research fraud pertaining to other aspects of the tamarin studies—including fabrication of data and false description of research methods—in the published 2002 paper and in other, ongoing work. Hauser ended up resigning from his Harvard post. It might seem that the study served, in the end, only to sharpen the conclusion of linguists like Chomsky. But instead, it led to significant research with another animal that would take the debate much further.

When University of California, San Diego, neurophysiologist Timothy Gentner reviewed the tamarin study (prior to the Hauser scandal), he immediately thought, *My birds could do this.* He had studied starlings for years, though he hadn't come to them through an interest in birds or birdwatching or ornithological science. He'd come to starlings through gibbons. Gentner found himself unemployed the summer after he finished college, and took a volunteer job as a docent at the Woodland Park Zoo in Seattle. There, he heard the daily vocalizing of the siamangs—large, black-furred gibbons native to Malaysia, Sumatra, and Thailand. They've been popular primates at the zoo for decades, famous for their singing. The males and females join in ritualized vocal duets culminating in the males' resonant boom, which can be heard, literally, a mile away, as those of us who have lived in the vicinity of the zoo can attest. Every day, the inhabitants of all the houses and apartments anywhere near the zoo have their dinner, their sleep, and their romantic tumbles interrupted (or enhanced) by the booming of the siamangs.

When Gentner started as a graduate student in psychology at Johns Hopkins, he couldn't get the siamangs out of his head; he wanted to study vocal patterning in animals and perhaps its implications for humans. His adviser suggested he start with the starlings they had in the lab, and Gentner never looked back. "The more I learned about these birds, the more interested I got," he told me. "And they have never ceased to amaze me. They are so adaptive, so clever. Every question I throw at them, they are able to answer, one way or another." He'd researched starlings for years by the time the tamarin study came around. He believed his starlings could recognize the patterns that the tamarins couldn't; in fact, he didn't think it would even be that hard for them. From his own observations, Gentner knew that both male and female starlings created wonderful and idiosyncratic songs by mingling an array of rattles, chirps, and warbles into uniquely patterned motifs and that they recognized one another as individuals based on these motifs. "Starlings are so tuned in to sequences. I knew they could do this. It was just a matter of figuring out how to test it."

Like Hauser, Gentner created a simple language, but his was based on starling-ish sounds—warbles and rattles. He tested eleven birds, and though it sometimes took thousands of trials, nine of the birds learned to recognize the recursive A(AB)B pattern more than 90 percent of the time, and with a high degree of proficiency, actually discerning the pattern when three new pairs of warbles and rattles were embedded between the original pair.

The implications were ripe for the picking. "Our research is a refutation of the canonical position that what makes human language unique is a singular ability to comprehend these kinds of patterns," Gentner declared. "If birds can learn these patterning rules, then their use does not explain the uniqueness of human language." That is, if a songbird could recognize recursive syntax patterns, then the place of recursion in Chomsky's linguistic views would have to be reconsidered.

And the wild rumpus began. Linguists, psychologists, biologists, ornithologists, evangelicals. Everyone jumped into the fray initiated by these tattered little birds—birds so unwanted by the wider world that Gentner didn't even need a wildlife permit for his research assistants to nab them from the trees. The not-yet-debunked Dr. Hauser's response to Gentner's study was seemingly open-minded. He pointed out that although the starlings could recognize recursive patterns, there was nothing to indicate that they could comprehend *meaning* in the patterns, which is true. (This, of course, would be much harder to study, since humans do not speak starling, or at least not yet.) But he recognized that Gentner's study went far beyond what he was able to show with the tamarins, acknowledged it as an "important paper," and claimed that he was inspired to think about ways he might re-create and improve his tamarin study by implementing some of Gentner's methods, perhaps using tamarin rather than human sounds and giving the monkeys more tries at listening to the short sentences.

Others, many of them linguists, disagreed strongly with Gentner's conclusion. When interviewed for the *New York Times,* Geoffrey Pullum, a linguist at the University of California, Santa Cruz, and co-author of *The Cambridge Grammar of the English Language*, said dismissively, "I'm not buying it." Pullum argued that the sentences involved in both the tamarin and starling studies were too simple to be of use in detecting cognitive abilities that had to do with grammar. It might have told us something about the abilities of starlings, but it said nothing about the more complex subject of human language. Yet as Gentner rightly pointed out on a UK radio program, "Primarily when humans are listening to acoustic patterns they're perceiving language." Here we are talking about the acoustic recognition of syntactical patterns, a linguistic ability, so if birds and humans can recognize the same patterns, why—and how—would we draw a line between them? And as a starting place for intellectual discourse, why would we? It seems far more expansive, more respectful of our complex intellect and the potential scientific secrets residing within the bodies and brains and lives of the creatures with whom we share the earth, to respond to such a study with our minds wide open, with a sense of curiosity and adventure.

Gentner described to me some of his new research that seeks to answer the question of why starlings might have the need for such advanced pattern-recognition capacities. In this work, he is focusing on the second part of the starling song, the one that mixes a starling's individually collected motifs into a long sequence. This section of the song has two components: the variety of motifs, and the patterns they occur in. One thing about starlings that is unusual among songbirds is that they are open-ended learners. Most songbirds learn their species-specific calls and songs in the first year, then they're done—they've successfully learned the "language" of their species. But starlings continue to learn new sounds, to bring in more sophisticated motifs and sequences throughout their lives. One morning recently I heard Carmen making an odd new sound. *YEEeeeEEEK!* Over and over. "Hi, honey." I offered one of her familiar phrases, hoping to distract her from this unpleasant noise, but she was not to be deterred. I finally figured out what it was when I was walking past Carmen's aviary into the dining room—she was perfectly mimicking the creak in our old oak floor.

Gentner was studying the way starlings of different ages pattern their motifs into sequences, and he discovered that older birds were more predictable than younger birds. The sequences of young birds are mostly random, while the sequences of individual older birds (a one-year-old versus a four-year-old, for example) are more predetermined. Say you are listening to a four-year-old male starling. If you have heard this bird sing before and paid attention to the sequence of his repertoire, then you can predict fairly well what the eighth (and ninth and tenth) motif will be. Females, it turns out, prefer such predictability (preference is measured easily enough by observing which males the females choose as mates). Older males make better mates by all agreed-upon bird measures: they declare and defend better and bigger territories, claim better nest sites, and, together with females, build better nests. They provide more and better food for their egg-sitting mates and their eventual chicks. And—the real measure of evolutionary-biological success—they fledge more young. Thus, life-or-death decisions for the starling are based on their pattern-recognition ability (and the female's leading part in mate choice feeds Gentner's intuition that females are slightly better than males at pattern recognition—something he has not yet been able to measure definitively).

But there is a point of diminishing returns—if a song is *too* predictable, it will become less attractive to the female. There appears to be a balance, a "sweet spot," Gentner suggests, between novelty and habituation. Too much novelty is unsettling; too much habituation is boring. Listening to Gentner talk about all of this, my mind turned to music, to the use of set motifs and refrains within a changing musical landscape—enough of both to keep listeners charmed and involved but still excited and waiting to see what comes next.

Just as writers use recursion to relieve the monotony of brick-by-brick sentences, musicians use recursion to allow their music to grow from within. The most famous example of recursion in beginning music-theory classes is Beethoven's Fifth Symphony, where the familiar *da-da-da-DUM* opening can be traced easily via recurrence and variation throughout the evolution of the movements. Mozart uses musical recursion frequently. Good music, like a successful starling song, represents a bounded complexity, and a lot of similar basic aesthetics are built into the patterning of language, human music, birdsong, and natural sound.

"I'm thinking of music," I told Dr. Gentner as we talked about the balance of predictability and variation in starling songs. "Yeah," he said, "me too," and launched into an esoteric exposition of early versus late Coltrane, during which I completely spaced out. I was pondering Imogen Cooper at the piano, wandering through variations on a theme in Mozart's Piano Concerto No. 17.

For the past few months we have had a quiet visitor in our household. Carmen's Aunt Trileigh and Uncle Rob (of the Guppy Incident) generously loaned us their Gnome Chomsky, the garden Noam statuette. Garden Gnome Chomsky was the first sculpture created by Steve Herrington for his small Portland, Oregon, company Just Say Gnome! Their mission: "To create and market garden gnomes . . . that will bring a bit of both humor and peace to people's lives and will hopefully also inspire deeper political, environmental, and spiritual awareness and reflection." We have loved having Gnome Noam here on the living-room mantel, emanating an aura of combined impishness and wisdom that captures the real Chomsky well. Like any good gnome statue, Garden Noam wears a fine red-pointed hat and brown boots, but instead of standing in a ring of mushrooms, he poses next to a table topped with scholarly books, wears glasses, and smiles to himself.

I have harbored a fantasy that Carmen would provide an irresistible photo op by perching on Garden Noam's little red hat. Dr. Chomsky is famous for refusing to make personal comments on certain issues, and I hoped that such an image might inspire him. I would send him the photos along with my linguistic queries, and he would be charmed into responding. Instead, Carmen was deathly afraid of Gnome Chomsky and wouldn't go anywhere near him for three days; after that, she'd tiptoe over to investigate only after I put dollops of her favorite peanut-butter-with-bits-of-arugula on his boots. Eventually she warmed to him, and as long as I stayed close by, she'd visit Garden Gnome and give his glasses little exploratory gapes. Contrary as ever, she has refused to sit on his hat for my imagined photo, but Tom managed to get a few nice shots, one of which I did send to Dr. Chomsky, who apparently remained unmoved. I never heard from him and must join the cadre of writers who are forced to say, "Noam Chomsky declined to comment." I love the photo anyway.

Carmen and Gnome Chomsky, the Garden Noam. *(Photograph by Tom Furtwangler)*

As it happened, Chomsky's rejection of the suggestion that Gentner's study might invite some re-thinking of his own work was immediate, complete, and rather truculent. For Chomsky, all the study showed was starling memory tricks, rattle counting, and number storage (all of which are rather interesting in a bird, let's not forget). "It has nothing remotely to do with language—probably just short-term memory," Chomsky said in his terse response to the *New York Times*. (Later research by Gentner uses more diversity of patterns and fewer repetitions to counter the possibility that the starlings are memorizing sounds rather than recognizing patterns.)

It's a bit of a sore subject for Chomsky. The starlings are not the first challenge to his theory that recursion is both unique to humans *and* universal to human language. There is a tiny tribe, the Pirahã (pronounced "pee-da-*ha*"), living at the mouth of the Amazon. There are only about three hundred and fifty Pirahã people in twenty or thirty small villages scattered along the Maici and Marmelos Rivers. They speak no outside language, and their own tribal tongue—a confounding mixture of clicks, rattles, lip flicks, air intakes, and birdsong-like bouts of prosody—have befuddled nearly every linguist to encounter the group, no matter how many years are given to its study. Surface stats on the language would lead to the conclusion that it ought to be simple to learn; there are only eight consonants (females use just seven) and three vowels. But the need for consonants falls away as tribal members slip into humming, whistling, and a complex rainbow of tones, stresses, and variations of syllabic length.

Daniel Everett, a linguist at Bentley University, is the only non-Pirahã in the world to speak Pirahã. He originally came to the Pirahã in the 1970s as an evangelist working for the Summer Institute of Linguistics, or SIL, an international group that trains language specialists to translate the Bible into the tongues of remote tribal groups. The organization doesn't attempt a lot of direct evangelization. Their belief is that once isolated communities get their hands on the Bible in their own language, their conversions will naturally follow.

In his decades of working with the Pirahã, Everett has slowly come to understand that as far as he can discern, they have no collective memory, no original creation mythology, and an insistent denial of complexity regarding numbers and amounts (the Pirahã recognize only *one, two,* and *many,* and attempts to teach them to count have failed). There is little grasp of or attention to the past or the future and little concept of the lives people are living when they are not standing there in the flesh, at least not in the way most of us typically understand such things. A person who walks into the woods to go hunting is simply "out of experience," says Everett. There is a profound "ethos of the present moment," as he calls it.

SIL's methodology was a failure. ("Have you *seen* this Christ?" the Pirahã people asked Everett.) And regarding language, Everett realized fairly early on that the Pirahã do not use recursion. Later studies by other linguists (including Hauser's research partner Tecumseh Fitch, who flew to the Amazon to do on-the-ground fieldwork) have failed to show that the Pirahã can recognize recursive patterns. In a new article, Everett stresses that this is not because the Pirahã lack intelligence; rather, it is "about the connection between their culture and grammar."

When he began his graduate studies, Everett was an enthusiastic disciple of Chomskyan linguistics, but the more he learned about the Pirahã culture and language, the more he fell away from his adherence to the notion of Universal Grammar. The Pirahã language, he came to insist, is a "severe counterexample" to UG, and he stressed his further belief that the Pirahã are not an isolated case but that scholars don't know of more exceptions because the entrenched linguistic theory has for so long stifled the impulse to inquire. "I think one of the reasons that we haven't found other groups like this," Everett said in an interview for the *New Yorker,* "is because we've been told, basically, that it's not possible." In the face of Everett's ongoing research, Chomsky continued to insist that "there is no coherent alternative" to UG.

Outside of linguistics, I follow Chomsky's political activism and commentary with interest. But I cannot mourn Universal Grammar's decline in the evolving field of modern linguistics. It is increasingly viewed as misguided and outworn at best and, some argue, unintentionally racist at worst. Obviously Chomsky would never have envisioned such a reading. But if the Pirahã do not recognize recursion, then making an absolute link between recursion and human language could certainly be interpreted as a diminishing of their humanity and that of any other tribal groups like them.

Gentner was not surprised by Chomsky's response to his starling studies, but he was disappointed that the negative reaction appeared to be based on personal attachment to a particular view rather than on stringent science. The starting process in Fitch, Hauser, and Chomsky's "The Faculty of Language" paper—the process of elimination that led the researchers to suggest the uniqueness of recursion—presumes a level of knowledge about what is happening in animal consciousness, communication, vocalization, and pattern recognition that is far beyond the human capacity to fully understand, both in 2002 when the paper was published and today. To presuppose that it is possible for three men to determine exactly what every earthly human group and animal species can or cannot accomplish in their communication is to begin from a place of astonishing hubris, an overstatement of what we do know, and surely also an overreach of what we *can* know.

No one is suggesting that human language is not unique and wonderful. As Gentner told me, "There will never be a nonhuman that can model human language, but until we can understand some mechanism that is shared across species, we can't even start to ask what is unique about human language." Yet we find that even among the most liberal intellectuals, there persists a notion that humans and human abilities must somehow remain at the center of the universe. The research on starling vocalizations and recursive pattern recognition pushes humans a little further off our self-created pedestal and deeper into the delightful mix of creatures that is the earth's truest symphony. We are at the edge of a new paradigm shift in the nature of scientific discourse, and we are being led by one of the world's most common, most reviled birds.

Darwin believed that all human capacities have an ancestral pathway, which of course makes evolutionary sense. There is no reason to tease out our consciousness in general, or our language in particular, from the wondrous, graced, earthen tangle in which we live. He wrote in *The Descent of Man,* "The sounds uttered by birds offer in several aspects the nearest analogy to language." That was 1871. Now researchers are observing that the pattern of vocal learning in human infants—from babbling to forming words to developing words into phrases and sentences—mirrors the way young birds learn their songs from adults. Duke University neuroscientist Erich Jarvis recently published an unprecedented study in the journal *Science* that mapped the genomes of forty-eight different species of birds. Jarvis had always been interested in avian voices, and determined years ago that the way birds learn song patterns seems to parallel the way humans learn to form words. He was hoping that his genetic research would expand on other work showing similar parts of human and avian brains are involved in vocal patterning. The results surprised even him. Jarvis and his co-researchers found fifty overlapping genes in humans and birds that correlate with vocal learning. In birds that were more adept at learning new songs, these genes were more often expressed. Jarvis's conclusions are momentous: "This means that vocal learning in birds and humans are more similar to each other for these genes in song and speech brain areas than other birds and primates are to them." There is at least one way that I am more biologically similar to little Carmen than I am to a chimp, the nearest animal relative of *Homo sapiens*.

The basic brain structure that we share with other animals, including birds, is ancient, predating complex communication. This suggests that the commonalities in our modern brains and genes around language are more likely to stem from convergent evolution—where two organisms evolve a similar physical characteristic in parallel—rather than from a close evolutionary relationship. But it's far easier to study birds long term in a lab than it is to study humans, and Jarvis hopes his and others' new work will shed light on the murky topic of language evolution. We have a fossil record to teach us about physical evolution, but we have no recordings of humans speaking seventy thousand years ago. If birds and humans learn language similarly *now,* it is possible that our evolutionary pathways to language have run much the same course.

While the reaction to Gentner's starling research from some linguists was negative, the response from the public was enthusiastic. His work was initially published in a scientific journal, but it got picked up by numerous public radio stations and was summarized in popular magazines and newspapers—more than usual for such a paper. This is not so surprising. I believe that it is a natural human tendency to seek and to *recognize* connection across species boundaries.

We are delighted when things we know to be true in our hearts and our bones are validated by science.

In 2012 an international consortium of prominent scientists signed a document called "The Cambridge Declaration on Consciousness," in which they proclaimed that animals, from birds to mammals to octopuses, possess consciousness similar to humans'. The paper stands at the forefront of what is said to be a new and more enlightened understanding. It is wonderful that animal consciousness is finally being recognized as respectable within high scientific discourse, and I hope that this paper and work like it will help ground a higher standard of ethics for animal treatment in science, agriculture, and entertainment. But I cannot help but think that as an academic declaration, it comes a bit late. Darwin made the same claim in print 162 years prior. And do most of us really need a scientific document to inform us that the animals we live with are conscious beings? I believe that the human sense of connection with the more-than-human world is innate and joyous. It is our truest way of being, of dwelling, of relating. It is not new; it is very old. It surfaces in the art and culture of every civilization across place and time—in stories of human-animal relationships that are based on respect, awareness, knowledge, and love.

I have no desire to confer on any animal a capacity that it doesn't have. There is no need. Animals have capacities enough—those we do understand, those we do not yet know, those we can never know because they reside in the unique minds of other-than-human beings. Starlings gather knowledge of their world by gaping. Parrots learn with their tongues, raccoons with the sensitive pads on the palms of their front paws, earthworms with their shining skins. "We lie in the lap of immense intelligence," wrote Emerson, "which makes us receivers of its truth and organs of its activity." And to me, this is the beauty of Gentner's work, and work like his. It reminds us of the creative awareness, at once scientific and poetic, that we stand on a continuum of being, of life. That we are part and parcel, along with every creature that crosses our path, of a fierce and beautiful intelligence.

Interlude

THE HEART OF TIME FOR BIRDS AND MOZART

Conductor Michel Swierczewski said of Mozart, "I have a theory that he is someone who lived faster than other people... When I think of his life as a film, I always see it as an accelerated movie." I have long held the same notion. By the time of Mozart's final illness and early passing, at age thirty-five, he had witnessed the death of his mother, his father, and four of his children, had suffered financial highs and lows, and had experienced the exuberance and anxious depression characteristic of artistic brilliance, all while composing quartets, quintets, symphonies, concertos, masses, dances, quadrilles, several of the finest operas ever written, and a beautiful requiem, the pages laid out across his lap as he died. Lessons, conducting, concerts, concertmastering, court composing. Thousands upon thousands of pages of manuscript written, copied over, published. Thousands of letters sent far and near. Thousands of miles traveled by horse-drawn carriage.

Certainly, in terms of the conditions of daily existence, Mozart was no different from the typical Viennese of his time. But in the pace of his work, juxtaposed with the hardship of his personal life and the complexity of his artist's mind, he was anything but typical. It is as if the space of a life opened wider in order to contain him.

New research confirms something that I have always believed to be possible. For smaller animals, time is perceived in slow motion. In a 2013 study published in *Animal Behaviour,* researchers at Trinity College in Dublin used flashing lights to determine the temporal resolution at which information can be processed by a variety of species. They learned that animals with smaller bodies and higher metabolisms (like houseflies or birds) perceive and process more information in a unit of time than larger animals with slower metabolisms (like elephants or humans). This is why, when we think we are being so sneaky with the rolled-up newspaper, the fly gets away most of the time—the sequence of events is unfolding more slowly for the fly. One commentator likened the effect to the "bullet time" sequence in the film *The Matrix,* where Keanu Reeves dodges the bullets coming at him in seeming slow motion while his coattails toss in the wind. To the shooter, the bullets are going at full speed, but in the Matrix, Reeves's character can duck the slow-moving bullets easily. According to this new study, birds live in the Matrix.

The implications of this research take us far beyond the notion of dog years—the idea that the lives of shorter-lived animals proceed at a pace that can be measured against a human time scale to find a comparable person-age. Instead, we are invited to think far more expansively and relativistically about time. I am not suggesting that a bird, say, with her fleet heart, experiences more in a short life of three years than we do in that same period but that her *actual perceived* life may be longer than three years. The measure is mysterious; the time of the bird's life expands beyond our typical calculation in ways that we cannot understand, at least not yet. Is it possible that some people, too, experience this time/space portal, allowing more experience to billow within and around them? That we can tot up the length of certain lives in our usual linear fashion but that these lives do not fit into this linear measure, that *more*, somehow, has been lived?

The potential for such experience is woven into our cultural mythologies. In the Western archive, we find Faerie, a realm into which one might enter through chance or mishap or, often, flute music. Faerie is the place where worlds meet—wild and domestic, dream and reality, language and poetry, human footsteps overlaid upon those of woodland creatures and leaves and mushrooms. It is the land of the imagination, and of the suspension of time, of the practical interlaced with the magical. Certainly it is the world of *The Magic Flute*.

Sometimes when Carmen sits on my shoulder, I close my eyes and listen. She weighs nothing; were it not for the tiny prick of her toenails I might not even know she was there. If all is quiet, and my ear is close enough to her warm feathered breast, I can hear her heartbeat. My heart rate, like Mozart's and most humans', is about 80 beats per minute.

Carmen's, like Star's and most songbirds', is about 450 beats per minute. Larger birds have slower heart rates (chickens' are about 245); smaller birds have faster ones (hummingbirds' are about 1,000). I put my hand on my heart and my ear to Carmen's breast and feel the pace of our two lives coursing by.

The metronome was patented by German inventor Johann Maelzel in 1815. All dedicated students of music are subject to its tyranny, but most composers continue to resist suggesting exact tempos for their work. Instead, the tempos at the top of a musical score are descriptive, suggestive, subjective, and highly relative. Allegretto (fairly brisk, but not fast), allegro non troppo (fast, but not too fast), lentissimo (slower than slow). We know that music can bend and change our perception of time, and myriad studies show how easy it is for humans to fall into a changed relationship with time when listening to skillfully composed music. The UK's Royal Automobile Club even determined the most dangerous music to listen to while driving: Wagner's "Ride of the Valkyries." Evidently, the risk is not that the listener becomes overinvolved in the music or that the music itself is too fast, but that the ecstatic nature of the music interrupts drivers' normal sense of speed, causing them to unconsciously drive faster.

The late Welsh poet John O'Donohue believed that "music is, perhaps, the art form that brings us closest to the eternal because it changes immediately and irreversibly the way we experience time." This, he felt, creates a bridge between the visible world and the invisible, or the *imaginal,* as it is called in the scholarship, where our usual measures make little sense. Mozart lived in this world of music, with ever-changing tempos in his head and a starling on his shoulder. Could it be that his experience of the passing of time was unique, that time unfolded for him in a distinctive, idiosyncratic way? I like to imagine that he experienced the interval of his own swift life with the expanded heart-time of a bird.

Eight

BIRDS OF A FEATHER

In June of 1787, Mozart entered a new composition, *Ein musikalischer Spass,* or *A Musical Joke,* into his catalog of completed works—the first piece he had finished since the death of his father and then his starling. The work is now designated K. 522, and is an unusual chamber ensemble for two French horns, two violins, viola, and bass. Mozart never shied away from complexity, but we know from his words and music that he was always against disharmony. This divertimento is a Mozartian anomaly, lurching wildly and unpredictably between keys and sprinkled with discordant accidentals. At the time of its publication, the piece was not given much regard as serious music, and those who paid it any mind believed that it was meant to parody the inepti-tude of the current musical establishment or was perhaps even a spoof of a particular composer. It is rumored that Czech composer Leopold Koželuch actually attacked Mozart on a visit to Prague, convinced that he was the parodied artist.

It has been a great sadness to me, and a kind of irony, that Carmen takes almost no interest in Mozart's music. I did, after all, pluck her from her nest and raise her from a scrawny, dying nestling to study a starling's relationship to the great composer. I wouldn't mind so much if she didn't seem interested in music at all, but she loves almost all other music. All except Mozart's—or most of Mozart's. As I mentioned, she does at least enjoy the final movement of Star's concerto. But there is one other Mozart composition that she dearly loves, and to me it seems another dimension of Mozart's *Joke*—as if he has reached beyond the grave to play one last trick. This starling I live with, this supposed kindred spirit to the maestro, sits blithely plucking at her feathers while Mozart's sublime Mass in C Minor drifts from the stereo speakers. But the much-maligned, musically dubious divertimento, the *Musical Joke*? Carmen leaps to life like an opera hero. She tilts her head. She looks at me as if something wild is happening and she expects that I, too, should recognize it. And if she is truly carried away, she will toss her little head back and sing. After that she will look at me again as if to say, *Was that good?* It's so cute and innocent it almost breaks my heart. I want to be annoyed, but instead I give her a kiss on the neck, which she hates. She fluffs the kiss off. "Very pretty," I tell her.

After Mozart's death, when his music was being passed around and played in his honor, musicians were reluctant to perform *A Musical Joke* because it made them sound incompetent; modern musicians are equally unenthusiastic. Some contemporary musicologists agree with the parody theory. Some feel that Wolfgang was working through his relationship with his father, who had died the previous month (though I find it doubtful that he would have treated this subject comically so soon after Leopold's death, or ever). Some think it contains a hidden message not yet understood. Some are convinced that this was an instance in which Mozart simply lost his way and created a superficial piece, lacking in significance. Most just find it unlistenable. Liner notes from a Deutsche Grammophon recording sum up the popular view: "In the first movement we hear the awkward, unproportioned, and illogical piecing together of uninspired material...the andante cantabile contains a grotesque cadenza, which goes on far too long and pretentiously ends with a comical deep pizzicato." By the end, the commentator declares, Mozart is writing like an "amateur composer" who "has lost all control of his incongruous mixture."

Such pronouncements suggest that the entire piece is a bizarre horror when in actuality there is much to enjoy—the composition on the whole is lighthearted and there are moments of bright energy and sweetness, especially the triplet-driven opening of the allegro, a section that makes me want to get up and twirl. If there seems to be a loss of control, as suggested by the Deutsche Grammophon author, I would argue it is a tightly controlled *appearance* of disorganization, which a true amateur could never pull off. Even so, Mozart didn't mean for the composition to represent his serious side. It was a playful piece with a riddle at its center: Where did this wild voice come from?

The description of this music should sound familiar. I have experimented with playing the most objectionable cadenzas from *A Musical Joke* alongside the recorded vocalizations of a starling who has lost himself in one of his long, rambling utterances and with Carmen's own singing. The melodies, or "unmelodies," of the quintet excerpts and the starling songs can be perfectly overlaid. I am not the first to notice the resemblance. When Meredith West and her husband, Andrew King, were pondering the subject of Mozart's starling while raising starlings of their own for research, the comparison became irresistible. The pair were students of animal behavior, not musicologists or Mozart historians, and with their fresh ears, they could detect something that centuries of musical commentary had overlooked. *A Musical Joke,* the researchers asserted in 1990, bears "the vocal autograph of the starling." They noted the fractured phrases, the tendency of starlings to respond to songs they have heard by singing them back off-key, to repeat parts that seem to have gone on long enough, and to delete parts that seem essential to the human ear. (One of the birds in West's study loved to mimic part of a song it heard in the household: *Way down upon the Swa—!* That was it. No matter how many times the starling heard the song, nothing could convince it to add *-nee River.*) On top of all this, starlings flexibly and unpredictably combine and recombine phrases, just as Mozart does in this piece, while tossing in starling-esque whistles and squeals. As Meredith West wrote in response to the Deutsche Grammophon commentator:

> "The illogical piecing together"—in keeping with the starlings' intertwining of whistled tunes. The "awkwardness" could be due to the starlings' tendencies to whistle off-key or to fracture musical phrases at unexpected points. The presence of drawn-out, wandering phrases of uncertain structure is characteristic of starling soliloquies. Finally, the abrupt end, as if the instruments had simply ceased to work, has the signature of starlings written all over it.

Revered ornithologist Luis Baptista of the California Academy of Sciences ordained the view. The piece that flummoxed musicologists, claimed Baptista, certainly does mimic the starling's innate vocal tendencies. He adds to the conversation by noting that the final cadence of the *Joke* is composed in a two-voice counterpoint—a Bach-ian element, but also a birdlike one. The structure of the syrinx (the avian version of the larynx) allows many songbirds, including starlings, to sing two and sometimes even more notes or tones at the same time (a technique perfected in the flute songs of thrushes). Baptista suggests that the counterpoint is another proof of starling influence, and I believe him. But I believe, too, that this cadence also represents the playfully joined voices of composer and bird.

As further substantiation of Star's influence on the sextet, we now know that *A Musical Joke* was composed in bits and pieces during the three-year period that Mozart lived with Star, and it was completed soon after the starling's death.

Thus, the piece morphs from musical error into musical eulogy—Mozart's idiosyncratic gift and tribute to the bird whose little life twined so thoroughly with his own. The affinity was an honest one. Bird and composer had much in common. Both maestro and starling shared an astonishing likeness in talents (mimicry, vocal play, musical gymnastics), personality (busy-ness, silliness, flirtatiousness, tomfoolery), and social priorities (attention-seeking!).

Like other members of his species, Star was flirtatious and prone to bursting into song, as was Mozart himself, who is said to have wandered eccentrically about, muttering in a musical recitative, a habit both operatic and birdlike. Like Star, Mozart was a gifted mimic. Certainly he could imitate any musical style in his compositions, and regularly did; he was commissioned to create particular styles of work for church or official occasions throughout his career. But he could also, even as a child, vocally imitate any of the prevailing opera seria styles in ranges from tenor to full soprano. When Mozart bragged in a letter to his father, "I can, as you know, pretty much adopt and imitate any form and style of composition," Leopold replied, "I know your capabilities. You can imitate anything."

As a child, Mozart was as wide-eared as a starling, swiftly learning new languages and absorbing musical styles and influences in his travels, and he would forever use modeling, imitation, and parody playfully and gorgeously in his own compositions. Always the impresario, Leopold urged Wolfgang to use this skill to professional advantage, and he did. But

Wolfgang also mimicked for fun, enlivening parties with his vocal and physical impressions of friends, other musicians, and even the emperor. He was surely surprised and delighted by the ever-changing, ever-mischievous vocal capacities of his pet, so much like his own. In their shared vocal play, their clever backing-and-forthing of aural possibility, Mozart found the closest thing to an avian kindred spirit that the green earth had to offer. A bird playmate evolved, it seems, just for him.

The other day, Claire and I were browsing at Pegasus, our wonderful neighborhood used bookstore, she at the Austen/Brontë shelf, I close by in Willa Cather, when I heard her whisper in teenage horror, "Mom, you have bird poop in your *hair*." I have lost count of the times she has had to make such proclamations in public, and they never cease to scandalize her. I've come to be rather philosophical about it. What can one do? Carmen sits on my head half the day, so naturally she poops there, and guano in the tresses is tricky business. While starling excrement wipes easily from a nonporous surface, the best way to get it out of your hair is to pick the solid bits off with tissue paper and let the rest dry, at which point it can easily be combed out. Usually, though, I forget about it, and it sort of blends in until the next shampoo.

When I spoke with Tim Gentner about his starling grammar research, the conversation came around to Carmen, and he was intrigued. Though he has worked with hundreds of captive starlings, he has never lived with one flying around his house. The things that usually surprise people about Carmen (that she's pretty, smart, tame, and can talk) are of course well known to Gentner. So I was amused by his question "Doesn't she make a mess?" By which he really meant: *Isn't there poop all over your house?*

Well, yes and no. Like all birds that do a lot of flying, starlings poop a lot. It's adaptive, essential in keeping their weight down for optimal aerodynamics. When a little waste accumulates in the intestines, they poop it right out to stay clean, light, and flight weight. Compare this strategy to a bird like a loon, say, or a cormorant—one that doesn't fly much but dives for a living and needs to be heavy and un-buoyant. These birds accumulate great stores of waste, which they eventually eliminate in shocking streams that float like rafts on the surface of the water; seeing them, you might think a whale had passed by. Though starlings defecate more frequently, their droppings are low volume. Most of the bird excreta we see on urban and suburban sidewalks come from birds like crows and pigeons—bigger birds with droppings that are substantial enough to stick around for a few hours or a day. But starlings and most other urban songbirds are smaller-bodied. Their poops blend in with the grass and the dirt, quickly wash away in a rain sprinkle, or are otherwise absorbed into the substrate.*

* This is not true, of course, in places where birds gather in numbers, such as colonial nest sites or autumn roosts, where the accumulated waste can be an eyesore and even a human-health concern.

The issue is more problematic for an indoor bird. Though I allow Carmen to fly freely as much as possible, she still spends a lot of time in her aviary and is there overnight, so that is where most of her droppings accumulate. I make sure that the newspapers lining the floor are changed every day and that her perches are scrubbed. As far as the rest of the house goes, recall that starlings are remarkably social. Carmen does not care to be flapping about the house by herself, exploring. She wants to be with us, *on* us. And since I am Carmen's primary caregiver, that's where her poop goes. On me. Or at least within arm's reach. I always have a square of tissue in hand, ready to wipe up any fresh pooplets that drop onto my computer screen or the floor or the book I am reading. I have an old shirt, now dubbed the poop shirt, which I pull over the top of whatever I'm wearing when I let Carmen out of her aviary. It's an imperfect system. I often forget to don the poop shirt, and when I take off my clothes to change into my pajamas at night, I discover that I have been wandering about all day with a healthy dollop of Carmen's doo-doo on the back of my sweater.

Friends or historians sometimes bring up the fact of random poop as an argument against my firm belief that Star was let out to fly freely in Mozart's study. Surely this would not have been allowed in a decorous eighteenth-century home, even one as eccentric as the Mozarts'? But a little starling poop here and there would not have fazed Mozart or his family in the least. Mozart was happily conversant with digestive matters. When Peter Hall, director of the play

Amadeus, spoke with Margaret Thatcher after a performance, he was bemused to discover that she was horrified by the maestro's vulgarity of manner as it was portrayed in this production. Evidently, the prime minister was not known as a patron of the dramatic arts; the premiere of *Amadeus* was the first time she had ventured to the theater in some fifteen years, and this because she was a fan not of theater, but of Mozart. Hall wrote of Thatcher's response to the performance in the introduction to a later edition of Peter Shaffer's play: "She was not pleased. In her best headmistress style, she gave me a severe wigging for putting on a play that depicted Mozart as a scatological imp with a love of four-letter words." Thatcher proclaimed it inconceivable that the man who wrote such "exquisite and elegant music" could be so foulmouthed. But while *Amadeus* does take many historical liberties, this is not one of them. Hall politely attempted to bring Mrs. Thatcher up to speed on the contents of Mozart's letters but was told, "I don't think you heard what I said. He couldn't have been like that." And Hall had no choice but to concede: "The Prime Minister insisted that I was wrong, so wrong I was."

Mozart reveled in toilet humor, a predilection he came by honestly; potty talk was a regular part of his upbringing. Naturally there was the usual playful poo-jesting between brother and sister, both at home and in letters while they were apart, but such humor continued long beyond childhood. Mozart's scatological jokes are often tossed off as part and parcel of the poor taste of adolescence, his overall immaturity, or perhaps even a kind of pathology. But excrement was a standard subject of conversation and joking between the seemingly straitlaced Leopold and the proper Anna Maria as well. When Wolfgang and his mother left for their European tour in 1777 (none of the family dreaming that Anna Maria would die in Paris), she assured her fretting husband in a letter along the way, "Don't worry, darling, everything will come right in the end. I wish you good night, my dear, but first, *shit in your bed and make it burst.*" (Emphasis mine.) This wording appears also in letters from the ostensibly prudish Leopold to his wife and son, and from Wolfgang to his first lover, cousin Maria Anna Thekla Mozart (the Bäsle, or "little cousin," as Wolfgang called her), whom he visited in Augsburg. As far as I have been able to determine, this is not a historical idiom but simply a lighthearted, and rather weird, family meme.

In a letter to the Bäsle, Wolfgang wrote, after a parade of other silliness, "Oui, oui, by love of my skin, I shit on your nose, so it runs down your chin." And further along: "I now wish you a good night, shit in your bed with all your might, sleep with peace on your mind, and try to kiss your own behind." And after pages of linguistic feats displaying a wild wit and questionable taste, the young maestro turned to fart jokes.

> *Now I must relate to you a sad story that happened just this minute. As I'm in the middle of my best writing, I hear a noise in the street. I stop writing—get up, go to the window—and—the noise is gone—I sit down again, start writing once more—I have barely*

written 10 *words when I hear the noise again—I rise—but as I rise, I can still hear something but very faint—it smells like something burning—wherever I go it stinks, when I look out the window, the smell goes away, when I turn my head back to the room, the smell comes back—finally My Mama says to me: I bet you let one go?—I don't think so, Mama. Yes, yes, I'm quite certain. I put it to the test, stick my finger in my ass, then put it to my nose, and—*Ecce Provatum *est!** *Mama was right! Now farewell, I kiss you* 10000 *times.*

When Constanze shared the Bäsle letters with Franz Niemetschek for his early biography, she wrote an accompanying note: "Although in dubious taste, the letters to his cousin are full of wit and deserve mentioning, although they cannot of course be published in their entirety."

I decidedly disagree with Hall's conclusion in his introduction to Shaffer's play that Mozart had an "infantile sense of humor" and "protected himself from maturity by indulging his childishness." Mozart's scatological bent is part of a wide-ranging wordplay that was bawdy and might have been immature in some ways, but in my view, the response of the entire Mozart family to an essential element of the human condition shows acceptance, wit, and a

* Correct Latin would have been *Ecce probatum est* ("There is proof").

sense of humor that is more indicative of playful intelligence than actual puerility.

Subscribers to the Sublime Mozart myth are naturally horrified, but in truth things may not be as bad as they seem. The Mozarts did not live in a whitewashed modern society—there was no water in the homes, no pipe systems, however rudimentary, no toilets of any kind. The facts of the body were close to the surface of everyday life. Nor was eighteenth-century Salzburg the laced-up theater of manners of the coming Victorian era; there was more room for casual jocularity.

And yet, even with such allowances, was the Mozart family's toilet humor—what would you say—*normal?* Probably not entirely. This talk is a bit coarse by any social standard, and for this reason, it offers a wonderful insight into the family. For two hundred years, we have unquestioningly swallowed the proffered image of Leopold as an uptight patriarch at the helm of a rigidly controlled family, and yet in the Mozarts' letters and ephemera and portraits, another image emerges. Here was a family that was intelligent and hardworking and concerned with status and success, yes, but also one that was comfortable together, that was jolly, silly, fun, and a bit raucous. All of us grew up visiting the houses of our friends and their parents, where we learned that there were families that made fart jokes and families that did not. The Mozarts were the sort that did.

And so, for Mozart, the finger-smeller, the shit-burster,

surely the tiny droppings that issued from Star, quickly wiped up, would have been nothing at all. Mozart composed on good paper, and it would have been easier to clean Star's droppings from his compositions in progress than it is for me to wipe them from my modern, poor-quality copy paper. Fair copies of manuscripts would have been kept away from the bird, just as I keep my beloved old leather diary, my complete Emily Dickinson, and other treasures out of arm's reach when Carmen is in my study (though they can stay out on the shelves and do not have to be put away completely—as I say, she doesn't venture much from my shoulder). Mozart's ink bottle would have had to be minded, but overall I expect Mozart had it easier with his bird than I do with mine. I am just waiting for my MacBook Pro to fizzle and short from fresh, liquidy Carmen poo sliding between the keys. When I took it into the shop because it was overheating and the fan was making a *thumpa-thumpa-thumpa* sound, the tech took it apart and showed me the "food" that was stuck in the mechanism. "Um, yeah," I confessed, "that's actually starling poop." With no apparent sense of humor at all, he brushed the fan clean and sold me a rubber keyboard cover.

Mozart's wordplay was not limited to bodily functions. In the above-quoted letter to the Bäsle, Wolfgang takes a flight of astonishing epistolary fancy. The whole letter becomes a

Spiel, a play, a playground. It does convey the kinds of thoughts and details of daily life typically found in a missive between dear young friends or lovers, but for the most part, it is a tumble of internal rhyming, verbal mirroring, echoes, synonyms, and puns that defy translation. It is nonsensical, but *calculated* nonsense, a sort of *Alice in Wonderland* letter, never mere blather, but a work of stunning, charming, goofy intelligence. From the greeting ("Dearest cozz bozz"), to the sign-off (his usual "I kiss you 10000 times," as he concluded almost every letter he ever wrote to his loved ones, though the number was sometimes one thousand or one hundred thousand or even a million), to the signature ("Old young Sauschwanz, Wolfgang Amadé Rosenkranz"*), there is never a dull moment in this long letter. "I have received reprieved your highly esteemed writing biting, and I have noted doted that my uncle gafuncle, my aunt slant, and you too are all well mell. We too, thank god, are in good fettle kettle. Today I got the letter setter from my Papa Haha safely into my paws claws."

Wolfgang makes fun of the Bäsle's declaration of affection in her recent letter: "You let it out, you expose yourself, you let yourself be heard, you give me notice, you declare yourself, you indicate to me, you bring me the news, you announce onto me, you state in broad daylight, you demand, you desire, you wish, you want, you like, you command that

* I am told that when properly pronounced, *Sauschwanz,* "pig's tail," rhymes with *Rosenkranz.*

I, too, should send you my Portrait." But then he gives in with flirtatious nonchalance. "Eh bien, I shall mail fail it for sure."

And while Mozart was at his freest in this note to his beloved cousin-lover, there is no shortage of linguistic acrobatics in the rest of his correspondence, and throughout his life. He slips so swiftly and agilely from his native German to riffs in English, Italian, French, and Latin, and tosses in his rhymes and puns so effortlessly that one could almost miss the fact that anything beyond the usual is going on. His most regular correspondent was his father, Leopold, who would worry and scold and demand news if he did not hear from Wolfgang frequently. Every letter, always, began with a rhyme, "Mon très cher père," and was signed with "I kiss your hands 1000 times". He wrote to Nannerl on the occasion of her marriage in 1784 (just a few months after Mozart moved with his family and starling to the Domgasse apartment): "Ma très chère soeur! Good gracious! It's high time for me to write to you if I want my letter to still reach you as a virgin! A few days later and—it's gone!" He sends congratulations and pleasantries and travel plans and concludes with explicit marital advice (which takes a feminist turn) in rhyming, rambling, iambic pentameter:

So, if your husband shows you cool reserve,
Which you feel you do not deserve,
But he, with knotted brow, thinks he's right:
Just tell yourself, well it's his way,

And say: yes, Master, thy will be done by day
*But my will shall be done at night.**

In February of 1786, Wolfgang attended a masked ball disguised as an Indian philosopher. He handed out pamphlets filled with clever riddles he had written himself, styled after the works of Iranian philosopher Zoroaster (the model for Nietzsche's Zarathustra and also probably for Sarastro in *Die Zauberflöte*). Wolfgang sent a copy to his father, and you can almost hear Leopold giggling with approval when he sent it along to Nannerl at her home in St. Gilgen: "The enclosure I'm sending you came from your brother. I solved the first 7 riddles right away just by reading them but the 8th is hard. These *fragments* are *good* and *true*. I suppose they are for moral edification . . . Please let me have the pamphlet back."**

---

* Sadly, Nannerl's marriage was never happy—it is doubtful that her wishes were ever much taken into account, day *or* night. Years before her marriage, she received a proposal from a fine young man whom it seems she truly loved, but Leopold objected to his trade and prospects and convinced Nannerl to refuse him. Leopold eventually urged his daughter into a marriage with magistrate Johann Baptist Franz, an older widower with five children and a home that was a loveless prison for Nannerl. She had three children of her own, including a son who was raised by Leopold. After Nannerl was married, she and Wolfgang slowly fell out of touch.

** An example of the high-spirited, political wordplay: "If you are a poor blockhead—become a *K _ _ _ _ _ _ _ r, Kleriker* [Cleric]. If you are a rich blockhead, become a landlord. If you are an aristocratic but poor blockhead, become whatever you can so you may gain your bread. But if you are a rich, aristocratic blockhead, become whatever you want to but not—I implore you—a man of reason."

Beyond all the chattering and wordplay, there is one similarity between Mozart and starlings that I would never have guessed. It is known that Mozart enjoyed drinking and was prone to tipsiness at parties. I cannot speak for Star, but wine is Carmen's favorite thing in her known world. She'll steal a sip any chance she can get, and we've discovered that she can even tell a wineglass from other vessels. If we line up a series of empty glasses on the counter—juice glass, water goblet, wineglass—she will fly straight to the wineglass and poke her bill down to the very bottom, seeking her sweet elixir and spreading her wings to keep her balance. If the cupboard is open for even a second, she zips straight to the wineglass shelf and knocks the flute nearest the edge to its death on the slate floor. Whenever we need to catch her to clip her toenails or accomplish some other starling-maintenance task, we just pour a little wine in the bottom of a glass and grab her when she is bill-down and tail-up. It's too easy. And though after such a stunt on our part, she will attempt to control herself and avoid wineglasses for a couple of days, in the long run it is too much for her. Her alcoholic tendencies overwhelm her suspicions regarding our motives. Occasionally I relent and share a little Cabernet from my own glass (just a bit—it's not good for her, and the last thing I need is a drunken starling flopping about). I love my vision of Mozart doing the same with Star.

Way back when I began exploring this story, I had no doubt that Mozart and his starling pet would find common ground in their adventurous vocalizing and musicality. But the more I have discovered of Mozart's personality and the more I learn about starlings by living with Carmen, the more I find the similarities between Mozart and Star to be more extreme than I'd ever dreamed: the unusual cleverness, the playful disobedience, the propensity for almost ceaseless chatter. Both were fluttering and curious and disorderly. Both were incapable of being still and quiet in a world so full of sound and happenings and beauty. Both shared the impulse to make wild, original, constant music.*

Beyond *A Musical Joke,* no one has suggested further direct connections between Mozart's starling and a particular piece of music. But that doesn't mean there aren't any. To the alert listener, I believe the lasting influence of Star on Mozart's work is everywhere in evidence. An infinity of birdlike phrases visit his later work. In arias composed for some of Mozart's rascally opera heroes, there are bouts of starling-esque mischief. And I do not think it is a simple coincidence that, after Star's death, a character appears in the Mozart canon that charmingly embodies all the shared personality quirks of both maestro and bird.

---

* It is said so often that Mozart was a "working stiff" (as one well-known essay about the maestro puts it) and composed only for work, for livelihood, for money. It is true that Mozart had to write for a living, and the bulk of his working life was carried out with an eye to this reality, but it is abundantly clear that he also wrote for fun, for art, and for love. He wrote for charity. He wrote silly songs for the entertainment of friends, and beautiful parlor music for the delight of guests. He wrote in his sleep. He wrote because he was Mozart, and music spilled forth from him, constantly and by nature.

Mozart composed much of his opera *Die Zauberflöte,* or *The Magic Flute,* in a tiny cottage set in the rose garden of Salzburg's Mirabell Palace, windows open to the sounds of birds and the cycles of nature that surface in the opera. This is the same garden where the von Trapp children twirled in the film *The Sound of Music,* wearing their curtain clothes and singing "Do Re Mi." The cottage still stands, but I was distressed to find it closed and padlocked for restoration when I was in Salzburg. I was still able to tiptoe around the cottage edges. After doing a little twirl of my own among the roses, I sat quietly and listened to the garden birds. These are the same species that Mozart heard as he composed (European robin, long-tailed tit, chaffinch, blackbird). I couldn't help whispering to them: "The souls of your ancestors still fly in the arias of *The Magic Flute.*"

Die Zauberflöte's ponderous libretto by Emanuel Schikaneder, with its Masonic overtones and dragging plot, is saved by two things: some of the most soaring operatic arias ever composed, and the comic presence of the character Papageno. Papageno appears in costume, a flurry of feathers—he is a bird catcher by trade, but in person more of an actual bird. In the opening scene, the opera's purported hero Tamino faints after being chased by a giant serpent. He is saved by the Queen of the Night's maidens, but when he awakens, only Papageno is there. "What are you?" Tamino wonders, noting Papageno's feathery garb and flittish demeanor. "Me?" Papageno is incredulous. "I'm a human being, like you," he replies with a sidelong glance that leaves us all in doubt.

When Tamino credits Papageno for his rescue, the bird catcher doesn't bother to correct him. The maidens return and punish Papageno for this falsehood by putting a padlock on his mouth. The worst of penances! Throughout the opera, there is a running gag that Papageno cannot keep from "chattering." The character of Papageno is social, charming, busy, strange, feathery, musically adept, unpredictable, troublesome yet delightful—it is not difficult to see where the inspiration for Papageno arose in Mozart's life. Papageno leaps about, unsure and aflutter, meaning well but getting into all manner of mischief. He is the Coyote-Trickster, the Shakespearean fool, his buffoonery masking his intelligence and capability. Everyone is in raptures over the supposed bravery of handsome Prince Tamino, but it is Papageno, not Tamino, who twice saves the heroine Pamina from being raped, and it is Papageno who finds the voice to join her in a powerful duet that sings love into existence.

The librettist Emanuel Schikaneder was a bit of a character. He had run a traveling theater and was himself a motley mix of impresario, playwright, player, and roustabout. In 1791 he settled in Vienna to run the Theater auf der Wieden, and here he renewed his long friendship with Mozart. It was Schikaneder's idea for them to collaborate on this new singspiel (or play with music—there are more speaking parts in *Die Zauberflöte* than in strict operas, where all speaking parts are sung in recitative). The librettist had a rich baritone and cast himself to play Papageno in the initial production. By all accounts, Schikaneder had a lively stage presence and doubtless made a wonderful bird catcher. Mozart kept the arias simple enough for Schikaneder's range, which was not as extensive as a professional opera singer's, but Papageno's melodies are so lovely, the sentiments so human and true, and the baritone voice so inherently gorgeous that listeners find nothing wanting.

Papageno is best loved for two arias: the first solo in which he announces himself as a bird catcher and the later ecstatic love duet with his mate, the equally feathery Papagena, during which the pair plan their future flock of little Papagenos and -genas. Both melodies are light, bouncy, silly, and enchanting, and along with the Queen of the Night's tormented rant, with its famous F above high C, these tunes from Papageno are the opera's zenith. Some musicologists have proposed that in these arias, Mozart is mocking the insipidity of the Viennese musical establishment. Or maybe they are meant to ridicule the compositions of Antonio Salieri, the lesser but at the time more socially established and respected composer. Or perhaps the figure of Papageno represents Salieri himself, painted as a halfwit. (The portrayal of Salieri in *Amadeus* is entertaining but fictitious—he did not poison Mozart.) I have made an effort to give credence to such theories, but in all honesty, it is difficult to believe that anyone who knows anything of Mozart and has truly experienced *Die Zauberflöte* could come to such conclusions. Within *The Magic Flute,* one message rises above all the others: Mozart loves Papageno. In one of the early performances, he even surprised Schikaneder by appearing offstage to play the sound of Papageno's glockenspiel himself!

Librettist Emanuel Schikaneder as Papageno. *(Engraving by Ignaz Alberti, 1791)*

Renowned Seattle Opera director Speight Jenkins (now emeritus) wrote an essay for the program of a 2011 production of *The Magic Flute* titled "Papageno's Magical Humanity," in which he claims that Papageno is the character in this opera with whom the audience can most immediately connect: "He wants a good life, enough food to eat, and above all a good wife. Mozart, from what we know of him, had the same feelings: He loved his wife, Constanze, and perhaps his most obvious connection in the libretto to his own feeling is Papageno's wanting Papagena to be his *Herzensweibchen* (wife of my heart), which was Mozart's pet name for Constanze." Mr. Jenkins has directed numerous productions of *The Magic Flute* and attended countless performances of the opera all over the world. When I asked him over a cup of tea at a Seattle coffee shop recently whether Papageno might represent some kind of derogatory comment by Mozart on the music of his day, Jenkins laughed. "Papageno is Mozart's Everyman," he responded. Then he paused and looked at me straight before saying, "Papageno is Mozart."

There are many persistent but untrue myths about Mozart—that he was a "man-child," always infantile; that he was broke his whole life; that he was buried in a pauper's grave. One of the most enduring is the notion, so commonly repeated in the Mozart scholarship, that he was the "most urban of composers," that he was at home only in the city and altogether isolated from the natural world in his life and his work. The misconception may be traceable to a work by Alfred Einstein (no verifiable relation to Albert), the music historian best known for completing the first thorough edit of Mozart's musical catalog (the Köchel catalog, still in use, which lists the works roughly in order of completion, thus the *K.* followed by a number that appears on each of Mozart's compositions). In his 1945 book *Mozart: His Character, His Work,* Einstein proclaimed that the composer had no sympathy for nature. The book was read by an elite audience in its time but would likely be nothing more than an obscure reference today if not for the fact that Thomas Mann, in the throes of his final illness, chose Einstein's volume as his deathbed reading. In a 1955 letter to his son Michael, the penultimate letter of his life, Mann wrote that he could sit up listening to music for perhaps an hour a day, but even this effort taxed his nerves. He would rather read Einstein's book. He swallowed Einstein whole and immortalized his untenable ideas with some of the last words that issued from his own pen:

> What especially interests me is that Mozart has no sense for nature at all, nor for architecture, or *Sehenswürdigkeiten* in general, but found stimulation only in music itself, and, so to speak, made music out of music, a kind of artistic inbreeding and filtered production—very curious.

The idea that Mozart's music is somehow sourced in music itself is philosophically intriguing, and there is no way to know how his genius unfolded in the mysterious process of composition, for which Wolfgang possessed an innate facility. But to make the leap from such musing to the notion that Mozart had "no sense for nature" is simply unsupportable.

Mozart did love the bustle of the city, and he was concerned about the social matters that came to the fore there—wealth, appearance, and especially acclaim. But even a cursory reading of Mozart's letters and knowledge of his daily activities show that he loved the natural world and turned to wild things for profound inspiration. Like Goethe, he was interested in recent scientific discoveries, and he paid attention to animals, weather, and the workings of the natural world. He accumulated a small gallery of fine bird prints, part of which was recently gathered for a temporary display at Mozarthaus, and his collection grew to include detailed plant drawings. He loved to be out of the city, to picnic with Constanze in wooded places, to wander at length with her in the Prater, the tree-lined, bird-filled park at the edge of Vienna. "I just can't make up my mind to go back to the city so early," he wrote to Leopold after a day out with his "pregnant little wife." And upon seeing the cottage in the Vienna woods where he was to be put up for a few days while visiting the chancellor, he exclaimed to his father, "The little house is not much, but the surroundings!—the woods—where he had a grotto built that looks as if it had been done by Nature herself—all of it is so Magnificent and so agreeable." Mozart did not conspicuously gush about his feeling for nature as the Romantics would, but his response was honest and heartfelt. He loved birds, loved animals, loved to ride his horse, loved to walk at the edge of the woods. All of this is expressed with joy in his letters and in his music.

Poet Gary Snyder wrote that wildness is "a quality of one's own consciousness," an elemental characteristic that ran deep in Mozart—he had a way of being, a habit of imagination that belonged in the realm of wildness and nature, regardless of where he lived. It is a quality that, at some level, we all share.

Carmen's domestic life is a trade-off. She doesn't have the freedom of a wild bird, but she doesn't have the perils of wild life to contend with either: exposure to extreme weather; the vagaries of food availability; competition with other starlings; parasites and diseases that social birds share; predators. And there are other perks to Carmen's life in our household. She is fed homegrown arugula by hand and has eggs hard-boiled just for her. She enjoys in-house cello concerts. She has a laptop and gets to watch *Seinfeld* reruns. She has a pet cat.

With all of this I sometimes forget Carmen's essential wildness, but then she will do something so completely and weirdly bird-ish that I am startled into remembering. Her sunbath is one of these things. On sunny days I grab a book and plunk myself in a chair by the bright kitchen window for this lazy-but-necessary starling health maintenance. Carmen perches on my arm or shoulder where the sun's warmth feels magnified through the window glass. She settles in, spreads her wings, tilts her head, opens her beak, and raises every feather on her body, as if she were a very soft porcupine. Bits of spittle form at the edges of her bill. To all the world, she would look extravagantly poisoned.

Like many birds, starlings enter a torpor-like state in the sun and spread themselves out so that as much light as possible can reach their epidermis. The many health benefits are believed to include vitamin D absorption, discouraging of parasites, release of oils that protect the skin, and possibly even a mental-health advantage—something akin to the restorative calm we humans feel when we lie on the beach or meditate.

Sunning birds look like dying zombies. Their pupils dilate, and they flop sideways on the ground. I cannot count how many times people have called me to describe a supposedly sick bird in their yard in just this pose. I've made sure that while Carmen always has some shade in her aviary, she also has a couple of places that get direct sunlight at certain times of day so she can sun herself when she wants to and needs to. But as per usual, she prefers company, and so she would rather sun on my shoulder or my arm, or on top of my head, just barely balancing on the tips of her outstretched wings. I marvel at the individual perfection of her feathers, all lifted one by one in this singular sunning ritual. And though Carmen sunbathes almost every day, I never stop wondering over the wild strangeness of it.

Carmen's water baths take almost as long and are even more trouble for me. I don't leave a dish of bathwater in her aviary because her splashing would make a mess with the bird waste and the newspapers that line the floor. If I put a

Zombie sunbath. *(Photograph by Tom Furtwangler)*

bowl in the sink, she will ignore it and stay on my shoulder—she wants me to *hold the bowl for her.* Baths are good for her, so I give in. Every day I hold her favorite turquoise Fiestaware bowl under the tap and let a thin stream of water fall and collect in the dish. Carmen jumps in and out several times, hopping from my wrist to the bowl, ducking her head and fluffing her wings over and over and splashing water all around the kitchen for as long as ten minutes. If I don't wear a raincoat, I have to change my soaking-wet clothes after. This might be a bit inconvenient some days, but it's also fun, and Carmen clearly loves her bath. After she finishes, she leaps to my shoulder and shakes like a dog, filling my ear with water, then flies straight to her aviary and is not the tiniest bit interested in coming out again for at least a couple of hours—she's busy preening every single freshly clean feather.*

Joyful water bath. *(Photograph by Tom Furtwangler)*

---

* Things took an unexpected turn when a certain Facebook video went viral: a muscular, tattooed man was giving his colorful little finch a bath in the sink, just as we do with Carmen, only instead of collecting water in a bowl, he used his cupped hands. Everyone on earth sent the video to me, and I had the seemingly innocent thought, *Wow, I bet Carmen would love this!* Mistake. It was tricky because my hands were much smaller than the guy's in the video and also because Carmen is much bigger than a finch. But she did love it and quickly became spoiled. For a long time she refused the bowl, and insisted on my hands, which got me even wetter.

When Carmen first feathered out, I thought I might have to *teach* her to take baths—to coax her into the water, maybe drizzle water over her head to inspire the ducking motion I'd seen in wild starlings during their sidewalk puddle baths. After all, I'd had to teach Claire to trust water and to wash her own hair when she was a toddler; perhaps young starlings learned to bathe by watching older starlings. Yet it turns out that the duck-lift-flap-splash motion is in her blood, and almost the second Carmen feathered out, she attempted to bathe in my water glass (while I was trying to drink from it). Like sunning, water baths are entirely innate. But I was soon to discover proof of an even deeper inborn wildness.

One day, Carmen was hanging out in her aviary on a branch by the window. All of a sudden, a Cooper's hawk perched briefly in the big camellia tree just outside, then threw himself, feet forward, into Carmen's closed window, talons aimed at her breast. Now, many birds flit at Carmen's window. There are chickadees, bushtits, house finches, and hummingbirds. Some of them, like crows, are almost as big as the Cooper's hawk. She observes them with a curious eye. Sometimes, she will hop over for a closer look; sometimes she will hop away and watch them, wondering, from a distance. This day, when a bird-eating predator she had never seen in her life came at her window, she let out a shriek I did not know she was capable of and hurled herself across the room and onto my shoulder, where she spent the next fifteen minutes panting and—another behavior I'd never witnessed—shivering. The Cooper's hawk banged against the window a couple more times (because he was flying from the nearby camellia, he was unable to get up much speed and was not at all injured, but he did appear to be exceedingly put out). He sat peering in from his perch among the pink flowers for two hours. Meanwhile, I had this suddenly wild bird on my shoulder who, in her own mind, had just narrowly escaped death. Here was a feral intelligence I hadn't been certain she possessed. But such evolved and wild knowing runs through blood, heart, and imagination—in birds, in each of us.

When I set out to follow the story of Mozart and his starling, I saw at its center a shining, irresistible paradox: one of the greatest and most loved composers in all of history was inspired by a common, despised starling. Now I muse upon the many facets of this tale, and it is wonderful, yes, even more wonderful than I had imagined. But looking back at the trail that I have wandered with these kindred birds—one in history and one in my home—I see also that, as both humans and birds so often are, I have been tricked by my attraction to the shiny little object. For in the end, it is not the exceptionality of this story that is the true wonder. It is its ordinariness.

In the creatures that intertwine with our lives, those we see daily and those that watch us from urban and wild places—from between branches and beneath leaves and under eaves and stairwells and culverts and the sides of walks and pathways—*we share everything*. We share breath, and biology, and blood. We share our needs for food and water and shelter. We share the imperative to mate and to give new life and to keep our young safe and warm and fed. We share susceptibility to disease and the potential to suffer and an inevitable frailty in the face of these things. We share a certain death. We share everything, constantly, every moment of every day and night, across eons. And in this shared earthly living, when we give our attention to it, we find the basis of our compassion, and of our empathy for other creatures.

And yet we have so much more in common than these of-the-body needs. We all poop, yes. But we all ponder, too, in a manner that may or may not be human but is whole and wondrous. We are at every moment surrounded by consciousness, a feast of unique intelligences. Every creature has its particular ways and wiles. Each being has its own presence, voice, silence, song, body, place. We are bound by our sameness and our uniqueness in equal measure—both spring from our shared being on a vital, conscious earth. *This is a wild communion*. And it is in this recognition that we move beyond simple compassion to a more certain, more essential sense of relatedness, of *kinship*.

Mozart felt this, I know. Like me, he was drawn at first to the shiny thing—in his case it was Star's singing back to him the song he himself had written. But in his elegy poem we see that a different relationship evolved. The bird's mimicry is not once mentioned. This is a poem to a kindred creature whose presence brought play, sound, song, joy, and friendliness to the maestro's life. And in the work that Star inspired, this is what we see too. A shared sense of mischief, music, and delight. The word *kinship* comes from Old English—*of the same kind*, and therefore related. *Kindly* and *kindness* also grow from this root—the bearing toward others that kinship inspires.

I have always thought of all creatures—all organisms, really—as relations. Whether wandering alone in deep wilderness or just leaning against a tree growing beside an urban sidewalk, I have had no difficulty feeling, as if in dreamtime, the roots of our relatedness—ecologically, yes, but also with an overlay of the sacred, the holy. Starlings, though pretty, were a rift in this vision. They fluttered outside this wholeness. But my thinking has evolved. Ecologically, it is true—starlings do not belong in this country, this city; but *relationally*, it is not true. We live together in a tangled complexity. I listen to the starlings mimic back to me my own profound ecological shortcomings. Carmen is a creature with a body, voice, and consciousness in the world. In this, we are sisters. And all these unwelcomed starlings on the grassy parking strip? Yes, they are my relations too.

The Cartesian belief in an absolute separateness of lives, bodies, and brains maintains a foothold in the traditions of our modern culture. We see it in the ways that we are pitted against one another in commerce, in education, and in the small, daily jealousies of our own minds. We see it in the ways that we continue to find it culturally acceptable to diminish animals in agriculture, in entertainment, and in scientific experimentation. And yet, when we are attentive, we find that we are not separate, not alone. We are not isolated little minds wandering on a large and indifferent earth. We are surrounded by our kin, by all of life, beings with whom we are wayfarers together. Instead of walking *upon,* we walk *within,* and this within-ness brings our imaginations to life. We are *inspired*—literally "breathed upon"—together.

Our creativity and our connection to other beings is tangled in a beautiful etymology. The words *creative* and *creature* spring from the same Latin root, *creare,* "to produce, to grow, to bring into existence." It was Ged, Ursula Le Guin's beloved young wizard of Earthsea, who learned after the fall of his individual pride that the wise person is "one who never sets himself apart from other living things, whether they have speech or not, and in later years he strove long to learn what can be learned, in silence, from the eyes of animals, the flight of birds, the great slow gestures of trees." Through such understanding we arrive at a new wholeness. We become more receptive and free in body and in imagination, and our unique potential for creative magnificence is enlivened. We become the listening artists of our own lives and culture.

And though, as Mozart learned with Star and I learned with Carmen, it is natural to come into attentive communion through the individual creatures before us—*this* bird, *this* raccoon, *this* tree (for we are learning that trees have their own systems of communication and knowing)—these individuals are not ends in themselves but a kind of window onto the totality of existence. I waited so eagerly for Carmen to mimic back the concerto's motif. Now I see that she has been calling out something much bigger, much more vital; she has been singing back the song of life, all of life, all the time.

Nine

MOZART'S EAR AND THE MUSIC OF THE SPHERES

In a dusty corner of Mozart's birth-house museum in Salzburg, there is a small room with lesser ephemera tucked into glass cases and tacked onto walls: cards and concert notes and silhouette portraits of distant relatives. I was there on a sunny day, and the rooms were dark. I'd been at the museum for hours, seen all I thought I wanted to see, and was feeling tempted to rush outside into the light and visit the aproned chestnut vendor I'd passed by the front step on my way in. But I reminded myself that I'd come all this way to contemplate such bits and bobs, so I sighed and stepped through the door frame. Nothing much caught my eye until I spotted a little lithograph, not much bigger than a postcard, at chest height near the door leading out of the room. On it, there were two penciled ears. The one on the right was labeled *Gewöhnliches Ohr* ("normal ear"). The one on the left was labeled *Mozart's Ohr.*

This sketch had been printed in the biography of Mozart by Georg Nikolaus von Nissen, Constanze's second husband, with the comment "The construction of Mozart's ears was completely different from the norm" and the claim that Wolfgang and Constanze's son Franz had inherited this rare ear shape. At first, both ears pictured looked normal to me, so I began to surreptitiously inspect the ears of other Mozart pilgrims in the room to get a better sense of what most ears really look like. Sure enough, the Mozart ear had at least two distinctions. The large, curved, upper part of the ear, called the antihelix (as I learned in my later research of outer ear anatomy), is broad, flat, and rather squared, and the small flap in front of the ear canal, called the tragus, is greatly reduced.

It is difficult to find portraits of Mozart to corroborate Nissen's claim; in most paintings, his hair is covering his ears. But a portrait of Franz and brother Carl as young boys (the only two Mozart children to survive beyond infancy) does seem to show a reduced tragus on Franz (his upper ear is not visible), and the book by Nissen was overseen by Constanze, who would have known more than anyone living about Mozart's ears. She insisted that certain things be whitewashed in the biography, but there would have been no reason to misinform readers about an ear; the sketch is likely credible.* It is probable that the strangeness of Mozart's ears caused him some embarrassment—maybe they are always covered in portraits

* Some scholars argue that only Mozart's left ear was atypical, since in a couple of portraits, his right ear is visible, though his left ear is always hidden. Nissen clearly refers to plural *ears*.

because he wanted to hide them. This auricular nonconformity remains rare today, but it is not unheard of and is colloquially called "Mozart ear" by specialists.

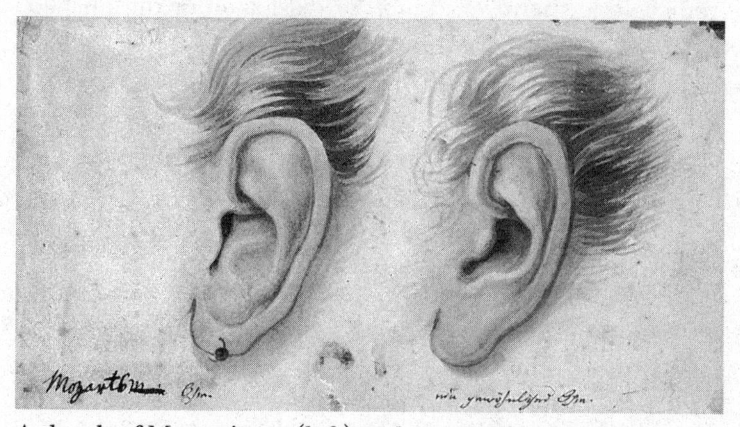

A sketch of Mozart's ear (left) and a normal ear.

What might it mean for a *composer,* of all people, to have an oddly shaped ear? A person like Mozart, whose life was dedicated to creation via sound? When I got home from Austria I talked with ear doctors and audiologists, and though everyone I spoke to emphasized that they were speculating, there was a general consensus. A fully formed tragus helps distinguish sounds that come from both the front and the rear. Sounds that originate in front of a person can sometimes be overwhelming to the inner ear, and it is believed the tragus provides a small barrier that subtly filters such sounds. Sounds that come from behind, by contrast, are already dampened by the ear pinna, so the tragus has an opposite function—it serves as a funnel that collects sound and bounces it into the inner ear. For someone with a small or nonexistent tragus, sounds coming from the front might be louder, fuller, more nuanced, or more distinct. Sounds from behind, however, might be dissipated or diminished, making it difficult for the listener to tell where they came from.

The impact of all this on Mozart as a composer is a matter of conjecture, but it does seem likely, at the very least, that Mozart experienced sound somewhat differently than most of us. It is impossible to say whether such an altered listening influenced his compositions, but it would certainly have led to an uncommon immersion in the aural dimension of life. I wonder how Mozart tilted his head to calm the sound before him, discern the sound from behind, or balance his heightened sense of the difference between the two. I think of Carmen's sweet turn of head—the way she lifts her strange bird ear whenever I call her name or when she is listening attentively to music or, really, to anything. Star did this too, as starlings do.

One morning early this spring I was walking in the wooded park near my home. The migratory flycatchers, tanagers, thrushes, and warblers were just beginning to arrive, thin from their long flights—all the way from Mexico, Central America, South America. I'd left my binoculars at home that day, intent on walking unencumbered and letting my other, nonvisual senses get in shape for the onslaught of growth and movement and birdsong that the season would bring. I heard a little chip in an aspen and stopped to listen.

There I was startled by a young man who passed, paused, and looked back at me. "I like the way you tilt your head," he told me, and walked on. The oddest compliment I'd ever received, but I liked it. *Yes,* I thought, *this is just the attitude I want to cultivate.* The tilt of the head, the listening for something just beyond normal hearing.

Birdsong is the perfect invitation to such listening. Eleanor Ratcliffe at the University of Surrey is engaged in a years-long study to determine the effects of birdsong on listeners. She is learning that human reactions to birdsongs are as varied as the songs themselves. Most of us are unsettled by raucous vocalizations, like those of a crow chasing a predator from its nest. But when we hear the cooing of pigeons, or the spring song of robins, or even just the background chatter of yard birds? Ratcliffe's study and those of others show that most people respond with decreased stress, increased calm, better concentration, brightened mood, and heightened creativity. Some find it easier to access meditative states while listening to birdsong.

All these responses are analogous to the way certain musical compositions lift and change our mood and increase our receptivity to the world around us. Both music and birdsong flit past our tympanic membranes, connect with our brains, brighten our minds, and transport our spirits. More than other ambient environmental sound, birdsong speaks in the musical language of pitch, rhythm, lilt, and repetition. But is it right to call birdsong music? By way of metaphor, there can be no argument. It is humans who chose the word *song* for the seasonal vocalizations of passerine birds, a word that is used in even the most academic of ornithological texts. But if we want to transcend metaphor and suggest that birdsong is music in the same way that human composition is music, then we are wandering into a scholarly fray that most of us did not know existed.

Birdsong carried through nearly every habitat on earth for millions of years before primates appeared, and so human evolution occurred against a backdrop of avian music. Cultures across all continents and times long before and after Mozart developed music that was inspired by and based on the sounds made by local birds. Over the decades, naturalists, ornithologists, musicologists, philosophers, and poets have found parallels and counterparts between the two. Scales, ornaments, trills, inversions, themes, variations. Not every passerine bird that sings uses all these attributes of human music, but all of them can be found in the combined repertoire of the world's songbirds. Darwin noted the resemblance of birdsong to musical composition and believed that birds possess an aesthetic sense. The well-known ornithologist Luis Baptista, in his paper "Why Birdsong Is Sometimes Like Music," writes, "Some birdsong is pitched to the same scale as Western music, which is one possible reason for human attraction to these sounds." Other prominent ornithologists note that white-crowned sparrows sing a perfect fourth interval between their first and second notes; that the canyon wren, with its gorgeous cascading series of notes that bounce against the desert stone walls, sings in the chromatic scale of twelve pitches per octave; that the wood thrush's layered song is pitched to the Western scale. The list could go on for pages.

One of the most intriguing comparisons of human music and birdsong was penned by the philosopher Charles Hartshorne, student of process theologian Alfred North Whitehead. Like Whitehead, Hartshorne saw the divine in the unfolding of earthly creation, in which humans participate. He was a gifted and committed amateur ornithologist, and he spent nearly an entire century—almost the whole of his 103-year life—immersed in the study of birdsong, research that culminated in his 1973 opus *Born to Sing: An Interpretation and World Survey of Birdsong.* It's an unusual book that combines scientific observation and quantification with the language of poetry, philosophy, and possibility. In his life of listening, Hartshorne located nearly every element of human musical composition in the songs of birds. *Accelerando, ritardando, crescendo, diminuendo.* Structure, rhythmic variation, melody, verse. The essential difference between avian and human music, suggested Hartshorne, is temporality, with the repeatable patterns in birdsong having an upper limit of about fifteen seconds (he evidently did not record starlings).

In spite of his well-documented examples and a book that is a wonder to read, Hartshorne has been criticized for being too expansive and speculative—too philosophical for the topic to hand, which some felt ought to remain securely in the scientific citadel. A recent paper published in *Animal*

Behaviour seeks to set the record straight. For his paper "Is Birdsong Music? Evaluating Harmonic Intervals in Songs of a Neotropical Songbird," Marcelo Araya-Salas, a PhD candidate in the animal behavior lab at New Mexico State University, studied the voice of the tiny nightingale wren, *Microcerculus philomela,* chosen for its complex, musical-sounding song. The bird is just a little bit bigger than a mouse, brown all over, with delicate curved black-and-cream speckling on its breast. Its eyes are huge and black. When it throws its head back to sing, it almost seems to shape-shift—the song is so startling, loud, and beautiful that the bird appears to become twice as large as it really is. When the song is finished, there again is just the avian mouse in place of the giant singing wren.

Araya-Salas gathered recordings of the nightingale wren from its entire range across Central and South America and analyzed songs from eighty-one individual birds alongside the well-known chromatic, major diatonic, and major pentatonic scales. These scales, he argues, are the ones that "represent the most intuitive pattern in which birds might base their songs." As a starting point for evaluation, he held up the "harmonic birdsong hypothesis," the notion that if birdsong was music, then consecutive notes in a bird's song should be closer to the harmonic intervals in these common scales than we would expect from chance. He evaluated 243 of his recordings alongside the musical scales to determine whether the wren songs conformed to harmonic intervals and determined that, in all of these comparisons, there were only six instances of wrens singing harmonic intervals, or about 2 percent—just what we would expect from chance, certainly not more. Araya-Salas argues, "If the frequencies are not determined by harmonic intervals in this species, it seems less likely that it happens in other birds with more complex song elements." He concludes that those who suggest a closer parallel between music and birdsong are simply misguided. "Documented musical properties in birds might be caused by cultural biases of the listener or misunderstanding of the physics of musical compositions."

Araya-Salas's conclusion gave me pause. *Documented musical properties in birds might be caused by cultural biases of the listener or misunderstanding of the physics of musical compositions.* In other words: *If you think birdsong is music, it is your own sad little misunderstanding.* The study is an interesting one with much food for thought, but the conclusions drastically overreach. The paper tells us that the song of one passerine species does not display harmonicity. That's one bird out of about four thousand species, each with a unique song. And the measure was just a tiny sample of Western musical scales—there are many other scales and harmonic forms people consider music, both in the West and across the globe. And the conclusion that this particular bird's song is not strictly analogous to a few particular Western musical scales? Well, we already knew that. But does that mean it's not music in some form or that it is not *musical?**

---

* Many respected scholars strongly disagree with Araya-Salas's conclusions. One of the most eloquent counterarguments comes from Emily Doolittle and Henrik Brumm in their paper "O Canto do Uirapuru: Consonant Intervals and Patterns in the Song of the Musician Wren,"

David Rothenberg is an academic philosopher, a musician, and a studied amateur ornithologist. In his wonderful book *Why Birds Sing,* he cautions us against failing to distinguish between what birdsong is *for* and what birdsong *is.* We know the ornithological explanations for the function of birdsong: the creation and defense of a territory, the declaration of sexual maturity, the attracting and securing of a mate. This is what birdsong is for. But as for what birdsong is? "Music," Rothenberg declares. In my college-level ornithology text, authors Joel Welty and Luis Baptista listed the usual reproductive, social, and individual functions of birdsong but added that they could not rule out the possibility that birds sing "from a sense of well-being" or simply "for the joy of it." And it was the philosopher Charles Hartshorne who added to the list of ornithological functions of birdsong "bliss."

The longer I ponder, the more I come to realize that the question of whether birdsong is strictly music, while a good question, is not *my* question. Determining a clear statement

---

published in the *Journal of Interdisciplinary Music Studies.* Here, the authors outline parallels between human conceptions of music and the song of a different South American wren, the aptly named musician wren, and many other songbirds as well. They make the welcome suggestion that explorations of such parallels are best approached from an interdisciplinary standpoint, with experts in music, ornithology, and audiology sharing their knowledge, to prevent one-sided, or simply mistaken, conclusions.

of the proper relationship between birdsong and music might have value for scientists and musicologists; in a strict academic setting bound by lexiconic definitions of words like *music* and *harmony,* the discussion gains a kind of interest and worth. This way of knowing has inherent beauty, and I do not mean to disparage it. But in this moment? With a Mozart quartet playing softly on the stereo as I write, and a tiny, round orange-crowned warbler in the tree outside my study window lending his trill to the top of the allegro movement? I close my eyes and hear them both entering my ears, and I cannot tease them completely apart. It would feel frugal and tightfisted to suggest that one is music and the other is not. And in such matters, I do not want to feel tightfisted. I want to feel profligate.

It is another day at the forested park near my home, and well into the summer. I am sitting in the grass, leaning against the thick trunk of a Douglas fir, and staring up into the branches of a thick and knotted old bigleaf maple. My mossy seat is not in a well-kept or frequently trodden area of the park. I was drawn by the raucous cawing of two juvenile ravens and wandered over for a look. Ravens are not common here, and this is the first time in some years that they have successfully nested in the park. Three young hatched, but one of these was killed while it was still floppy and small, most likely by an off-leash dog.

I am being eaten alive by biting flies, but I stay where I

am because one of the young birds has flown from the tangled woods and into the tree right next to me, from where she calls to her sibling. *Craaaww, craaawww!* So loud! This young bird is perfecting the scream with an underlying low croak that characterizes the vocalizations of her species, already distinguishable from any other bird. But in spite of the ruckus she is making, the little raven appears calm. Quiet in spirit, loud in voice. She looks down at me, unfazed, as naive young birds are. She fluffs her wings, calls again. It is good to be in her company. In my notebook, I sketch the tree, sketch the bird. I squint at my drawing, and even then it is not good. I close my eyes and listen.

Crow. Red-breasted nuthatch. The chirring of starlings in a nearby nest, the tapping of a downy woodpecker. Robins. The spiral song of the Swainson's thrush. Human children. Traffic. A motorcycle. The low of a ferry coming in to dock. Bushtits, chickadees. Olive-sided flycatcher. A bird whose voice I do not know (which makes me vaguely uncomfortable). A rustle on the ground in the tangle of ferns and huckleberry and nootka rose. A mouse? A Pacific wren? At this urban/woodland park it is just as likely to be a rat. A woman calls to her child, "Gaabbbyyyy!"—the only sound louder than the ravens. Spotted towhee. More crows, mad ones (they don't like the ravens, but there may also be a sharp-shinned hawk nearby). I keep my eyes closed as all the sounds soften, yet heighten and blend. I visualize them surrounding me, a map-blanket of resonance all around me. But now there is a scritch right next to my ear, so startling that I cheat and peek. A brown creeper hanging on the tree bark! I can hardly keep calm. The creeper pays me no mind in my stillness and the tiny scrape of creeper toenail in my ears is as loud as a truck. I want to squeal, but I close my eyes back down, farther down. I calm my excited creeper nerves. There is my breathing and, can I hear it? Yes, there is my heartbeat.

When I finally open my eyes, I am Rip Van Winkle, unsure of how much time has passed. All the sound is still here, still everywhere. And the raven is still above me. "Hi," I whisper. But her eyes, now, are lightly closed. Perhaps she too hears her own heartbeat, faster than mine. Surely she does. But here is her sibling raven flapping in the branches. Clumsy and new, he flies to another tree, perhaps a hundred yards away, and *craawws* loudly. This is too far; the siblings want to be close. My raven looks up, shakes off her own young silence, screams, and flies in the direction of her brother. There is a movement of grass, of bodies. The sounds continue, all of them and more. They rise, fall, become almost silent, rise again. "Is this it?" I wonder out loud—why not? The rest of the earth seems to be wondering aloud. Am I wrong to think that this is the music of the spheres?*

---

* The brown creeper is a tiny songbird with woodpecker-like adaptations for creeping up barky tree trunks as it searches for insects. It is dark brown, like bark, but with a bright white belly, a slender curved bill, and a stiff tail to support it vertically. It is one of my all-time-favorite birds and was also beloved by the late social and environmental activist Hazel Wolf. I met Hazel when I began working at the Seattle Audubon Society years ago, where she was the secretary until her death at almost 101. I got in the habit of meeting her at her simple apartment in Seattle's Capitol Hill neighborhood and taking her to lunch every month or so. We would walk to the little Thai restaurant down the block. The walk took longer if it happened to be election time—Hazel would make me join her in leaving flyers on every door we passed to "get out the vote." She was grassroots, old school, and formidable. Almost every lunch, she would tell me about how

When we use this phrase colloquially, we refer to a poetic sensibility or a philosophic notion. To "hear the music of the spheres" is to feel in harmony with life, with self. We have seen something of incomparable beauty or achieved a state of bliss through meditation or yoga or mountain climbing. We find ourselves ridiculously alive after a horrific accident, or we have fallen suddenly in love. Somehow, the world itself is singing to us, and we are listening in a rarefied way.

Our ancestors were more attuned to the movements of the heavens than most of us are today. This was partly because the night skies were more visible due to the absence of bright lights, but people who depended on gathering, hunting, and small-scale farming for survival also relied on the movements of the stars to locate themselves in time, in seasons, in the cycles of planting and harvesting. In Egypt, the appearance of shining Sirius presaged the annual flooding of the Nile. The regular movements of the stars above were reassuring and gave some sense of rhythm and predictability in a time that was plagued by a harsh awareness of the ephemerality of life. Against the predictability of the stars, though, lay the discomfiting wandering of the planets (the word itself is

---

elated she felt the first time she'd seen a brown creeper, a memory that circled in her elder's mind and one that I never tired of hearing. The brown creeper forages by clinging to the trunk of a tree with its feet and making its way up. Up, up, up. It doesn't skirt up and down like a woodpecker. Just up. When it's ready, the creeper flies down, and then begins its journey ever upward again. "Just like me," Hazel would say, twinkling.

from the Greek *planetes*, "wanderers"). If the stars had meaning in relation to earthly cycles and events, as they clearly did, then the planets must have messages as well. But these messages were incomprehensible, impossible to read (when an elite priesthood claimed that they could decipher the messages of the planets, they became the first astrologers). Humans' pursuit of meaning in the planets likely began two hundred thousand years ago or more.

We can only wonder over the speculations of the early sky watchers. But by the time Pythagoras was born, in 570 BC, on the Greek island of Samos, there was in place a strong human desire to find a fundamental regularity to the bafflingly complex planetary movements. Pythagoras himself is a shadowy figure, a kind of cipher about whom little is known, and yet somehow the history of geometry has come to be written on his thin biography. It is possible that even his eponymous theorem might have been outlined centuries earlier by mathematicians in Egypt, India, or Babylonia. But we do know with certainty that Pythagoras and his followers sought links between numbers, geometry, and the structure of the natural world, believing that all of these were interconnected and that the connections themselves had not just mathematical but also ethical and spiritual significance.

In their search for mathematical and earthly harmony, the Pythagoreans explored music and musical instruments. They recognized that when you pluck two strings, one twice the length of the other, a perfect octave is produced. Exploring further, they found that strings with a length ratio of two to three produce an interval called a fifth, which is pleasing to the human ear; Western music was largely developed around such intervals (the violin has four strings, each a fifth apart). Though the Pythagoreans did not possess the mathematics or the astronomical knowledge to posit anything but the vaguest of theories, their melodic explorations led them to propose a "music of the spheres," a harmonious relationship that linked the planets one to another in just the way strings arranged on a violin produce notes that vibrate in harmony.

The idea of a universal harmony founded in sound is not unique to Western mathematics. In Hinduism there is the notion of *shabad,* which is sometimes interpreted as an audible life stream. Humans can enter this current through the chanting of the universal syllable *om,* which activates the energy centers in the human body and promotes physical and spiritual resonance with the wider universe. And in the Gospel according to John, 1:1, it is written, "In the beginning was the Word," which many scholars claim would be more accurately translated as "In the beginning was the Sound." Not a primordial soup, but a primordial hum.

Two thousand years after Pythagoras, the German mathematician Johannes Kepler, an advocate of Copernicus's new heliocentric theories, followed up on the notion. Kepler sought to understand the sacred architecture of the solar system and to find mathematical harmony in the way that planets are placed in space. After a series of misguided propositions, Kepler eventually became the first known mathematician to accurately describe planetary orbits. He formulated the principle now referred to as Kepler's first law of planetary motion, which states that orbits are not circular, as was initially presumed, but elliptical; and later his second law, which established that the spread of the ellipse increases the farther a planet is from the sun. With these laws, Kepler gave us a roundabout pathway back to Pythagoras and a planetary orchestra.

Pythagoras based his inquiries into harmony on musical notes, but as any string player knows, no note that is actually performed is mathematically pure—every note played on an instrument is influenced by other notes, the subtle vibrations of the other strings, and even nearby instruments (when my daughter plays a D on her cello, the D string on my violin a few feet away hums, even though no one is touching it). These are musical overtones, and they create the complexity we look for in a good musical instrument. Geophysicist David Waltham elucidates this concept in his book *Lucky Planet*. The overtones, he stresses, are what make instruments unique. If a pure sound were possible, then all instruments would sound very much alike, and the orchestra pit would be dull indeed. Instead, overtones allow for the unique sounds of a violin versus a piano versus a French horn, creating a world of interaction between material substance, space, and sound.

The simplification might torture a cosmologist, but there is an analogy in the movement of the planets. The orbit of each planet in our solar system has a characteristic wobbling, a tilt that, as it oscillates, interacts with the oscillations of the other planets' tilted, slightly wobbly, ever so slowly changing orbits. The resulting vibrations are the *planetary* overtones, and when they mingle, they create a complex hum, the background music of our universe, our planet, our lives. Simple, perfectly parallel spherical orbits would not allow the friction that could create such a sound, so Kepler really was on the path that led to our modern understanding of the music of the spheres.

This planetary hum is fifty octaves below the hearing range of humans, but scientists have modeled the orbital oscillations and raised the frequency to human range that they can play back for us. The sound is a kind of chaotic scream that, as it goes on, lowers in pitch and increases in volume, as would be expected of expanding elliptical orbits. (Waltham again suggests musical instruments by way of analogy: A cello is lower and louder than a violin of the same shape.) The sound, writes Waltham, must be more like the familiar chaos of an orchestra tuning up—not wholly unpleasant, but still discordant and unpredictable—than of the orchestra coming together to perform a concerto. More like a gull, he says, than a nightingale. Or perhaps, I cannot help thinking, like the whistle of a starling.

And overlaid upon this music of the spheres, this present-but-unhearable starling-esque hum? There is the music that we hear, and make, daily. There is the music of Bach, of Messiaen, of Mozart. Of the woodland wren, the woodpecker's drum. The laughter of an infant, the drag of an old woman's cane against the sidewalk. David Rothenberg wrote: "Species are supposed to sing only for their own kind, but the more I listen, the less I'm sure. Yesterday I heard all the thrushes singing together in the laurel woods: veeries, hermits, wood thrushes. They all seemed to be getting at one total song." I recognize this as the song from my woodland meditation. Human and natural music both make mischief with our ingrained patterns of separateness.

Years ago, I heard the late Zen Buddhist teacher Robert Aitken give a reading at Elliott Bay Book Company, Seattle's famed indie shop. After, those who had gathered for the talk walked out the doors and into a warm late-summer evening. Aitken Roshi had ended with a short meditation; our minds were hushed and open. It seemed sudden when someone pointed and called out, "Look at the birds!" There in the urban sky was a cloud of starlings, thousands of starlings, swirling in one of their great, mesmerizing orbs. "It's a sign," someone whispered.

And it was a sign. Autumn was coming. In the spring and summer, starlings divide into couples and then family groups in the business of mating and nesting. But in the late summer, this exclusivity breaks down, and starlings begin to gather into the groups that will grow into their huge fall and winter flocks. Flocking has many benefits—through force of numbers, birds find and share food, roosts, and warmth, and, in flight, they foil aerial predators. When a hawk sees a solitary bird, it can focus on that bird as prey with efficient single-mindedness. But a flock becomes an organism in itself, making it difficult for a predator to zero in on an individual bird.

Starlings seem to take the evolutionary mechanism of flocking into the realm of high art when they gather in hundreds, thousands, sometimes even a million birds and turn about the sky in mysterious, graceful, spellbinding dance-clouds called murmurations. These are evasive maneuvers, complicated enough to beguile even a peregrine falcon, the most formidable of aerial predators. But they bewitch the human mind as well. As we watch them converge into great spheres, then swirl into funnels and ellipses, starling murmurations lift us into an elated, almost hypnotic state.

Murmuration. *(Photograph by Donald Macauley)*

Some ornithologists claim that starling flocks were originally named murmurations because of the varied songs and sounds that starlings create. But starlings do not call out much during their flock-dances, and I am more inclined to agree with other avian etymologists who believe that the name comes from the whisper of wings, so many wings, together in flight. Beneath a murmuration, I feel that I am kneeling in an ancient cathedral that ought to be silent but instead whispers overhead with the gathered prayers of hundreds of years of pilgrims. But here is a much greater cathedral—the entire sky—and the prayers are the light brushings of feathers.

For centuries, humans have looked up at these murmuring flock-prayers and asked, "How?" How do they spin and change and rise and gather and spread, all of them, at exactly the same time? With so many birds in the air, it is imperative that everyone keeps going the same direction—if even a single bird goes its own way, there will be aerial crashes and broken wings. Yet there has never been a satisfactory explanation for the phenomenon. Perhaps there was a lead bird that all of the rest watched and followed? But sometimes the flocks spread the length of several city blocks—there was no way that the bird at one end could see the bird at the other. Maybe the leader gave a vocal signal, unhearable to us on the ground? Maybe they were involved in a Rupert Sheldrake-ian group mind, a kind of morphic resonance? Maybe so. The explanation for such preternatural coordination seemed to transcend traditional biology. And now that the technology has caught up enough to give researchers the ability to look carefully at murmurations through high-powered video, high-resolution slow motion, and computational modeling, this seems to be true—starling flocks do fly beyond the parameters of biology and into the realm of cutting-edge physics.

In 2010, University of Rome theoretical physicist Giorgio Parisi published a paper in the *Proceedings of the National Academy of Sciences*. Parisi and his team found that starling murmurations function in the same way as various natural systems on the cusp of a shift, described by scientists as "critical transitions." These are systems that are poised to tip into a sheer transformation, like metals becoming magnetized, liquids turning to gas, or gathered snow in the moment before an avalanche. In such transitional moments, the movement or velocity of one particle affects all the others, no matter how many others there are, in what is called a "scale-free correlation." In terms of starling murmurations, the change in velocity or movement of one starling would affect all the rest, whether there were fifty others, or fifty thousand. But if every bird is busy paying attention to all the others, how does one individual launch a change in the whole? And how does that whole respond so quickly?

The answers remain murky, but a couple of years after their initial paper, Parisi's team took their research further, paying attention to the correlation of each individual to the birds closest to it. It turns out that the change in the movement of one bird will affect the *seven* birds closest to it. Those seven birds will each affect seven more birds, and their movements will ripple, scaling rapidly, through the flock. How it happens so incredibly fast is still a mystery and the subject of ongoing research, but it is postulated that these moments of transition in starling movements may mirror universal principles at work in the proteins and neurons that underlie the makeup and movement of all creatures. Thus, starling murmurations might be the most visible and also the most winsome iteration of biophysical criticality, a mirror into deeper, unseen, all-embracing secrets of life that have yet to be understood. Watching them, I feel that mystery viscerally; I feel my head swirl and my body sway. I always thought this was because the movements of murmurations are so graceful, and surely that is part of it. But it may also arise from an unconscious identification with the same movements at work within the neurons of my own brain and neural synapses. Deep calling unto deep, as the psalmist sang.

The late Welsh poet John O'Donohue spoke of the earthen wisdom possessed by our animal brothers and sisters. "The animals are more ancient than us," he urges. "They enjoy a seamless presence—a lyrical unity with the earth. Animals live outside in the wind, in the waters, in the mountains, and in the clay. The knowing of the earth is in them." And by our attunement, by the tilt of our heads, our enlarged listening, we enter this knowing in a state of shared awareness and being that O'Donohue refers to as "interflow."

It is from this beautiful, feral place that we are able to respond to the breath of inspiration that summons us to the fullness of our creativity. Full, because we are cognizant that we are not a lone pair of hands or eyes or a single voice, that we do not create in isolation but bring our gift, the art of our lives, to one another, to the earth. We each touch the seven starlings closest to us in our own murmuration, and the ripple spreads faster than we could have imagined. We create from what Clarissa Pinkola Estés calls the Rio Abajo

Rio, the "river beneath the river." The song just beneath our typical hearing, the murmuration that calls the tiniest neurons of our brains into flight.

And what is this wild summons? What art is asked of us? The gift offered is different for each but all are equal in grandeur. To paint, draw, dance, compose. To write songs, poems, letters, diaries, prayers. To set a violet on the sill; stitch a quilt; bake bread; plant marigolds, beans, apple trees. To follow the track of the forest elk, the neighborhood coyote, the cupboard mouse. To open the windows, air the beds, sweep clean the corners. To hold the child's hand, listen to the vagrant's story, paint the elder friend's fingernails a delightful shade of pink while wrapped in a blanket she knit with the deft young fingers of her past. To wander paths, nibble purslane, notice spiders. To be rained upon. To listen with changed ears and sing back what we hear.

Finale

THREE FUNERALS AND A FLIGHT OF FANCY

While writing this book, I felt constantly superstitious about Carmen's well-being. Whenever an animal has a book written about it, that animal ends up dead. Dewey the Library Cat. Marley the dog. Wesley the owl. Mumble, the owl who liked sitting on Caesar. Mij, the otter in *Ring of Bright Water.* There was a sweet little book penned in the 1980s called *Arnie the Darling Starling;* Arnie succumbed to a foot infection in the last chapter. Writing about an animal seems to be the very kiss of death. I worried throughout the project that this chapter would end up being titled "Four Funerals."

Carmen did everything she could to assist in the fulfilling of this grim prophecy. I've listed many of the common ways for household starlings to die; Carmen constantly thinks up new ones. It seems she tries to find a way to kill herself every single day. Besides getting herself locked in the refrigerator, her many attempts have included:

—Flying headlong into a closed window, after which she lay motionless on the floor for fifteen minutes while I knelt next to her whispering, "You're going to be okay, little one."

—Ingesting a rubber band that I had to eke out of her esophagus with surgical delicacy so it wouldn't become tangled in her intestines.

—Attempting to peck at the nose of Delilah the cat, whom we forgot to lock up before letting Carmen out of the aviary.

—Climbing into a narrow cellophane bag that she found who knows where, getting her wings pinned, and thrashing helplessly about until discovered, by which time she was gasping for air and the bag was filled with the fog of her breath.

—Trying to eat something too big to swallow, then almost choking on it; a raisin, an almond, a garbanzo bean, a snap pea. And, once, a whole grape.

Carmen's Near-Death Magnum Opus was flying through an open window. Of course we are normally religious about closing all the windows when Carmen is flying loose in the house. But this was a sunny day; Tom was home alone, listening to Bob Marley and dancing joyfully around the kitchen. He forgot all about the open window above the stove, and when he let Carmen out to join him, she flew straight through it and into the backyard. Tom ran out the door just in time to see her wing around the house and, it seemed, into the open world. He wandered the streets, calling her most familiar words—"Hi, Carmen! C'mere!" But nothing, nothing, nothing. By the time I arrived home, an hour later, Tom was in a panic, huddled in the backyard with his head in his hands. "I lost Carmen," he told me, near tears. Meanwhile, a gaggle of teenage girls was arriving to meet Claire for a pre-high-school-dance beautifying party. Then the pizza guy showed up and wanted money. All was utter mayhem.

The prognosis for starlings who have been raised indoors and escape into the world is not optimistic. They are not experienced at feeding themselves; they don't know how to behave in starling society, and so do not have the protection of a flock; they think cats are their friends. Lost starlings have not learned the geography of their neighborhoods as they would had they been raised among flying outdoor birds, so if they go exploring, they cannot find their way home, even if they want to. I learned on the website Starling Talk that some lost starlings have been recovered after they tamely approach nice people who realize they must be pets and put up FOUND BIRD signs, but this seemed a long shot. I was certain that Carmen was lost forever.

After searching and calling all over the neighborhood for a couple of hours, I finally heard a loud starling contact call from the very top of the thirty-foot cypress in front of our house. Starlings regularly use this high-pitched chirp to stay in touch with their flock or with their young. I didn't know if Carmen knew how to make this vocalization, and I couldn't see the bird that the call was coming from, so very high in the tree. But I hoped. We borrowed a tall ladder from our neighbor and placed it against the side of the house, close to the cypress. Tom climbed up and up, thinking that if the bird we were hearing was in fact Carmen, she would see him. "Hi, Carmen!" he kept calling. *Great,* I thought, fretting. Now I had to worry that I would never see Carmen again *and* that my husband would plunge to his death from the top of a precarious two-story-high ladder. But the bird in the tree began to make its way down, branch by branch. "Carmen!" I called. It was her. As soon as she saw Tom, she flew to his shoulder and clung breathlessly to his T-shirt. She was as happy to be back as we were to have her.

As I dot the *i*'s on this manuscript, Carmen almost miraculously continues to flourish. Maybe starlings really *do* have nine thousand lives. But in this story, there are still three funerals to tend to.

All three funerals—father, son, and starling—have been widely misunderstood in the Mozart mythology. Leopold's came first. In the spring of 1787, Wolfgang heard from Leopold that he was very ill. Leopold was a hypochondriac, over-inclined to seek sympathy, and Wolfgang knew this, but his was a ruminating heart, and he would still have worried endlessly. Then a new letter: Leopold was much improved! Then word from a friend: Leopold was ill indeed, perhaps mortally. Poor Wolfgang! He was not prepared to lose his father. But in this moment, he gathered his strength and his quill and composed one of the most famous letters in the Mozart epistolary canon. The letter began roundaboutly but typically for a missive between Wolfgang and Leopold. Mozart called up the unique intimacy between father and son as he wrote about the world of current music and, within it, one of their favorite subjects—others' declining talents. This, he whispered to Leopold, is "*just between you and me.*"

> *Ramm and the 2 Fischers—that is, the bass singer and the oboist from London—were here during the Lenten Season; if the latter did not play any better when we heard him in Holland than he is playing now, he certainly does not deserve the reputation he has.... To put it in one word, he is playing like a Miserable student.... Well, it's the Truth—*

Eventually, Wolfgang was ready to come around to his worries.

> *This very moment I have received some news that distresses me very much—this all the more as I gathered from your last letter that you were, thank God, doing very well;—but now I hear that you are really ill! I need*

not tell you how much I am longing to hear some reassuring news from you yourself; and, indeed, I confidently expect such news—although I have made it a habit to imagine the worst in all situations.

There is no reason to doubt the sincerity of Mozart's distress, and the self-knowledge revealed in the last sentence is as touching as it is true. He never recovered from the death of his mother, never lived down his self-blame, never for an instant ceased to worry about Constanze and all his loved ones. Wolfgang himself was ill at the time with acute kidney problems. But in this moment, instead of further fretting, he penned a short meditation to Leopold, a kind of philosophy of death and dying. It was a message of consolation, meant equally, as I read it, for father and son.

Death, if we think about it soberly, is the true and ultimate purpose of our life. I have over the last several years formed such a knowing relationship with this true and best friend of humankind that his image holds nothing terrifying for me anymore; instead it holds much that is soothing and consoling! And I thank my god that he has blessed me with the insight, you know what I mean, which makes it possible for me to perceive death as the key to our ultimate happiness.—I never lie down at night without thinking that perhaps, as young as I am, I will not live to see another day—and yet no one who knows me can say that I am morose or dejected in company—and for this blessing I thank my Creator *every day and sincerely wish the same blessing for All my fellow human beings.**

Mozart expressed himself honestly, and his sentiments surely served as a quiet gesture of both penitence and forgiveness between father and son. It is a beautiful letter. The passage is often quoted as an expression of Mozart's feelings toward death and a suggestion that his own passing, just four years later, was, if not actually welcomed, at least met with equanimity. It is read as an overlay to the more sublime moments of the *Requiem*. But I think it is important to remember that this is the expression of a twenty-nine-year-old emotionally distraught musical genius who is about to find himself parentless. In my reading, this passage is a cry to God from a son outwardly expressing calm while inwardly on his knees, rending his garments. On Mozart's desk was the draft of *Don Giovanni,* with its flawed Everyman whom no critic has ever been able to fully condemn, redeemed as he is, at least on some level, by the harmony of Mozart's arias. The don is about to be claimed by the flames of hell. Mozart was more conflicted than the lines of this letter would indicate, yes, and yet surely he repeated the sentiments to himself as, awaiting more news of his father, he lay

* Many of the ideas about death expressed in this letter are inspired by Freemasonry, and Leopold would have been familiar with its premises, having joined Wolfgang's own lodge in Vienna. Another influence was likely the thought of philosopher Moses Mendelssohn (grandfather of composer Felix Mendelssohn). A copy of his tract *Phaidon, or The Immortality of the Soul* was well thumbed by the time it was listed in Mozart's effects after his death.

down before a fitful sleep. *Death: soothing, consoling.* This was the last known letter from Mozart to his father.

Leopold died on May 28, 1787. He was sixty-eight years old, a good age for the time. It is generally accepted that Mozart's failure to attend Leopold's funeral in Salzburg was a protest, conscious or not, against the passive-aggressive authority and control that Leopold wielded over Mozart for the whole of his life. The psychology of the relationship was certainly complex and damaging, but Mozart never outgrew a feeling of honest devotion to his father—a mixed sense of love, guilt, and unfulfilled obligation that would plague him until his own death. It is surely true that Wolfgang could not face the death of his father straight on. But at the time Leopold died, Constanze was immobilized with a septic leg, there were young children in the house, and the couple was deeply in debt. Mozart could not leave his wife, could not afford the travel, could not face the expectation, once in Salzburg, of laying out cash for mourners and services. He did not protest the funeral; he simply could not go.

Star died just two months after Leopold. I've discussed that, as a tribute to his starling, Mozart arranged a formal funeral, invited friends as formal mourners, and performed a dramatic reading of the elegy he had composed for the bird. This is the whole poem:

A little fool lies here
Whom I held dear—
A starling in the prime
Of his brief time,
Whose doom it was to drain
Death's bitter pain.
Thinking of this, my heart
Is riven apart.
Oh reader! Shed a tear,
You also, here.
He was not naughty, quite,
But gay and bright,
And under all his brag
A foolish wag.
This no one can gainsay
And I will lay
That he is now on high,
And from the sky,
Praises me without pay
In his friendly way.
Yet unaware that death
Has choked his breath,
And thoughtless of the one
Whose rime is thus well done.

Reveling in the contrast is irresistible: Mozart absented himself from his father's funeral, then buried a common bird with pomp and flute music and original poetry! Many believe that the starling's funeral was simply a farce—one of Mozart's many indecorous social shenanigans. Others suggest that the starling service provided Mozart an avenue for catharsis in the face of his father's death and perhaps also a transference of duty; he was doing for the starling what he believed he ought to have done for Leopold. I find truth in both these views. A fancy bird funeral certainly would have appealed to Mozart's sense of the absurd. Meanwhile, for all Mozart's complicated genius, his personal relationships were touched with a childlike simplicity and a deep neediness. Feeling that he had neglected Leopold, Wolfgang surely found comfort in the ritual of a funeral, a receptacle for his displaced grief.

But there is a third way of thinking of the funeral, one that would be obvious to anyone who has lived with a starling. While honoring the psychological nuance of the theories above, we can also recognize that Mozart felt honest sorrow over the loss of his bird. In the three years he lived with Star, Mozart had struggled for professional recognition; faced periods of financial despair; become estranged from his sister Nannerl, the bosom confidante of his youth; and lost two beloved children and his father. Through all of this, the starling had been present as an always-cheerful, ever-mischievous companion, a mirror of what was liveliest and most creative in Mozart's own soul, a simple, constant friend. Frivolous rhyming cannot mask the truths contained in this poem. Mozart knew the funeral was silly and over the top. But it was also sincere, an act of affection, a parting gift.

He buried his bird in the garden and marked the tiny grave with a stone.

Physically, Wolfgang was not a strong specimen, and he knew it. He was small—probably just over five feet—and slender. In the fine unfinished oil painting by Joseph Lange (Mozart's brother-in-law, husband of Constanze's sister Aloysia) on display in Salzburg, Mozart is pensive, with soft cheeks and the suggestion of bags beneath his eyes. The painting shows just his face, and one can imagine such a head topping a doughy body, but it was not so. Mozart was always thin. He had never been in robust health, had suffered recurring episodes of rheumatic fever as a child, and was bedridden time and again with respiratory ailments, some of them life-threatening. He survived bouts of scarlet fever, acute infectious polyarthritis, and a juvenile case of smallpox that left him pockmarked for life. In the years preceding his death, Mozart was often bedridden with severe colic and several other unidentifiable illnesses. In spite of his assurances in the death letter to his father that he was always serene in public, he suffered from the headaches, melancholia, and anxiety to which profoundly creative minds are often prone. And he was, like his father, a bit of a hypochondriac. Even with all of this, at age thirty-five, with a desk and mind full of unfinished compositions, with a docket of travel to performances of successful work, and with a family that was not yet comfortably provided for, Mozart was unprepared for his final illness, which caught him unawares.

Unfinished portrait of Mozart. *(Joseph Lange, 1782)*

In July of 1791, before the onset of his sickness, Mozart was approached by an anonymous stranger and engaged to compose a requiem for the wife of an illustrious Viennese gentleman. The commission was generous; Mozart accepted, began the work, then set it aside to complete *La clemenza di Tito* for the coronation of Emperor Leopold II in Prague.

It sounds apocryphal. Mozart, unknowingly near his own death, is drawn into a prescient, ghostly commission by a shadowy figure who in some biographies wears a dark hood. And yet it is true. The stranger was an emissary of Count

Franz von Walsegg, whose wife, Anna, had been only twenty-two years old when she'd died suddenly in February. (The seeming Victorian-Gothic depiction of the emissary in the story—gaunt, hooded, and shadowy—was actually an accurate description of Walsegg's courier Franz Anton Leitgeb, who was tall and thin with a taciturn disposition and a dark Turkish complexion and who always dressed in gray.) It is likely that Mozart knew both the count and his young wife, as Walsegg often invited the Viennese cognoscenti to his country home for music. One of the reasons Mozart was approached anonymously might have to do with the count's proclivity to commission music and then to pass the work off as his own. When guests would ask about the composer of a piece that had been performed in his parlor, Walsegg is known to have asked them to guess. And when the response was a polite "Well, it sounds like it could have come from your own quill," the count would just fix a pleased little smile on his lips. Passing off a requiem by the great Mozart as his own tribute to his beloved dead wife was the probable motive for secrecy.

Mozart gloried in recognition, but he needed the commission and likely believed that the truth would eventually come out. Indeed, by the time the completed *Requiem* was delivered, after Mozart's death, everyone knew who it was for and who had written it. The count is said to have been angry but was perhaps placated by the fact that the composition increased in value with Mozart's death.

While at work on the *Requiem* in November of 1791, Mozart became sick quite suddenly. His symptoms included swelling of the hands and feet, high fever, rash, and intense sweating. His body became so bloated and painful that he could barely move. Constanze provided light cambric cloth to her sister Sophie, who sewed a new dressing gown that Wolfgang could put on without sitting up—he just held out his arms so the robe could be slipped on and tied behind his neck with a ribbon. Even in such a state, he claimed to be delighted with his sister-in-law's clever industry and declared himself happy to wear the new gown.

In his last weeks, Mozart worked on the composition from his sickbed, poignantly aware that he was penning a requiem as the potential imminence of his own funeral mass became more and more real to him. It is not overly romantic to believe that this gives the work much of its power and urgency. There are passages of the *Requiem* that might be overlaid with sections of Mozart's death letter to his father—sweet, restrained harmonies that welcome death as a consoling friend. But these are just moments. The music rises and twists into darker moods and fearsome crescendos. The music of the *Requiem* is never ugly. It proclaims Mozart's foundational faith in beauty and harmony while embracing gloom, shadow, even fright, with forthrightness. He is more honest in his music than he could ever be in his letters.

And again it sounds apocryphal, but it is nevertheless true that Mozart continued to work with all the dwindling strength of his mind, body, and spirit to complete the *Requiem* as he lay dying. He could not have helped but recognize (and weep over) the symbolic significance, and yet his motive was in part practical. When the work was done, Constanze could collect the handsome second half of the commission. He worked with the guilt of a man who knew he should have done more.*

Mozart could not sit fully upright; the score was spread all over his bed, across his bloated belly. On the working manuscript, his last notes were scrawled in an ill and quivering hand. Unlike the scene in the movie *Amadeus,* Salieri was not there taking dictation on how the *Requiem* should be completed. It was Wolfgang's good friend the composer Franz Xaver Süssmayr who finished the work after Mozart's death, with the maestro's instruction in his head and in a passably Mozartian manner. Naturally, purists criticize—there are errors of harmony and an occasionally glaring mismatch of style. Even so, most modern conductors prefer Süssmayr's attempt over those that came after.

Mozart died on the fifth of December. His symptoms were consistent with rheumatic fever, but a deadly streptococcal infection was sweeping Vienna at the time, and modern scholarship suggests that this is what killed him, perhaps

* Though Mozart himself had composed pro bono works to fund the Freemasons' organization in support of the widows of composers and artists, he had failed to subscribe for his own soon-to-be-widow. Without foundation, and with overt antifeminist sentiment, Constanze has been vilified for centuries by Mozart biographers and fans as dull, unmusical, under-devoted, moneygrubbing, and a generally undeserving partner to Mozart's genius. But after Wolfgang's death, Constanze worked against all manner of difficulty to settle her husband's debts and to face the practical task of establishing financial stability for her little family: she appealed to the emperor for a widow's pension in light of Mozart's service at the court; she organized concerts of Mozart's work and oversaw the publication of his music. Over time, she and the boys became financially secure.

preying on a system already weakened by the fever. It is almost certain that the medical treatment of the day hastened the hour of death. Mozart was let for over two liters of blood in the days just before his passing, and immediately before he died, he was wrapped in cold compresses.* Constanze's younger sister Sophie had argued that the compresses would be a shock to her brother-in-law, now delirious and barely conscious. They were applied by the doctor over her objections, and Mozart died soon after. Sophie's bedside recollections are the only extant first-person account of the event, and they have the ring of truth, though her remembrance that Mozart died with the timpani part of the *Requiem* upon his lips seems a stretch. The physical details at the time of death—the severity of the bloating, the stench, the discoloration and projectile vomiting—all of this was suppressed for years, believed to be unfitting to the reputation of a great composer. But listening to the *Requiem*, we can find all these things. Ugly, feared, redeemed in the whole.

The fact that Mozart was buried in a common grave and at a service with no mourners present has long been taken as a disparaging comment upon musical Vienna—a society that

* The tools of the bloodletting treatment of Mozart's time are on display in the collection of mismatched artifacts on the top floor of the Mozarthaus museum. When I'd read about bloodletting in the past, I had always imagined a small razor, but instead here was a horrific brass box full of thick, hinged, hook-ended blades, never sterilized, of course, that looked like instruments of torture.

first overlooked and then spat upon its most gifted composer at the hour of his death. But the circumstances of Mozart's funeral were entirely in line for those of the middle class after the reforms of young Emperor Joseph II, who, in a frenzy of enlightened rationalism, worked to overturn the extravagance of previous generations, invoking a sense of practical simplicity and propriety in funerary matters. Bodies of respectable citizens were laid out in the church (St. Stephen's Cathedral, in Mozart's case), then wrapped in a linen shroud and, within two days of passing, transported in a reusable coffin to a graveyard outside the central city and there buried in a common grave of six to twelve bodies, each doused with lime to prevent stench and the spread of disease. Mozart himself was a supporter of such reforms.*

The Mozarts were not well off at the time of Wolfgang's death, but any notion of a "pauper's burial" comes from a misunderstanding of the customs of the very particular time

* It is usually suggested that Mozart was buried only in a linen shroud, but of all the Josephian reforms, the edict to allow only shrouds, no coffins, for burial was met with the most public outcry. It went one step too far. Several months before Mozart's passing, Joseph reluctantly lifted this element of his edict and allowed families, for a fee, to purchase a simple coffin that would be deposited in the common graves. Volkmar Braunbehrens notes in his detailed *Mozart in Vienna* that in the list of Mozart's funeral expenses, there is a line item for the hearse. Hearses were provided for free unless a coffin was purchased—evidence, argues Braunbehrens, that Mozart might have been buried in a coffin. He notes also that Georg Nikolaus von Nissen, Constanze's second husband and one of Mozart's early biographers, wrote vaguely that "the coffin was deposited in a common grave and every other expense avoided." But Nissen wasn't there; he had Constanze as a source, but she did not make the funeral arrangements, and it is possible that he just assumed there was a coffin. There is for now no definitive answer.

and place. Myth has been heaped upon myth. In 1856 the *Vienna Morgen-Post* published an excerpt from a memoir by someone named Joseph Deiner, a man who claimed he was present at the funeral. The author had a dramatic winter tempest rain down on Mozart's small funeral cortège.

> The night of Mozart's death was dark and stormy; at the funeral, too, it began to rage and storm. Rain and snow fell at the same time, as if Nature wanted to shew her anger with the great composer's contemporaries, who had turned out extremely sparsely for his burial.

In fact, records from the time report mild weather and just an occasional light mist on the day of the funeral, but the tale of a poignant storm was gleefully adopted by biographers and remains strong in the popular imagination.

To say that Vienna forgot Mozart or metaphorically spat upon him in the manner of funerary arrangements is to ignore the gracious outpouring of attention in his honor that marked the following days and months in Vienna and beyond. There were notices in the papers across Europe. On December 10, Emanuel Schikaneder joined others to arrange a funeral Mass in St. Michael's Church near the Hofburg, where completed portions of Mozart's *Requiem* were performed. In Prague, there was a grand memorial service with a full orchestra and chorus, and there were said to be several thousand in attendance.

Like any good Mozart pilgrim in Vienna, I slated an entire day for the exploration of St. Marx Cemetery, where Mozart was buried two days after he died. When I was confirming directions at the tourist information center (I am an obsessive confirmer during travel), the gentleman behind the desk told me I would take the 71 tram outside the central city to the Landstrasse District. He wrote this down and handed me the paper with a little smile that struck me as odd. The people of Vienna are helpful and friendly, but restrained, not at all smiley. When I was buying tickets to a local music performance, the ticket seller switched from German to English to talk to me before I'd said a word. "How did you know I was American?" I asked. "Because you smiled at me," he said. He laughed and added, "Don't worry, it's nice. We don't mind." Later a volunteer docent at the tram museum explained the tourist information worker's smirk. Joseph II's reform required that all burials take place beyond the city's main wall for sanitary purposes, and this cemetery was created to fit this edict, opened as a burial place in 1784. For the last hundred years, the tram line from the city center to the cemetery has been the same—the 71. To "take the 71" is a colloquialism for dying, sort of like "kick the bucket."

It was a cool, sunny October morning. I was in love with Vienna; I was in love with Mozart; I scoffed at my jet lag; I couldn't wait to board this tram. It took me past ornate government buildings, through fancy neighborhoods, into less-fancy neighborhoods, past tract houses, and eventually to a vast industrial area that looked dark and gray, even in the sun. Here the tram stopped, and the driver, whom I'd pestered about letting me know where to get off, glanced at me and tipped his head toward the open door. "Cemetery?" I asked. He nodded; other passengers nodded. I slowly disembarked. The stop was in the middle of a busy highway with traffic rushing both ways on either side of the bus shelter. I tried to orient my tourist's map and decided to head uphill, though every direction would take me along a speeding, dirty, multilane expressway. I followed the local people, who seemed to know how to cross the street without dying, consulted my map again, and walked tentatively northward. Soon there was no one. Who would walk here? It was just me and cars.

After about fifteen minutes of walking, wondering, and consoling myself with the uplifting traveler's perspective that I was, at least, *somewhere,* if not where I'd hoped to be, a bald man dressed all in biker-style black came toward me. I mustered my courage. *"Sprechen Sie Englisch?"* "Leetle," he answered, lifting his thumb and forefinger to show me the measure of just how little, in fact, he knew. No light showed between them. "St. Marx Cemetery?" I ventured. "This St. Marx!" He was jubilant. He pointed to the big green highway overpass sign, and indeed it read ST. MARX. Hmm. "Cemetery?" He looked confused. "Dead people?" I offered. Nothing. "Dead." I dropped sideways as if dying. I let my tongue loll out of my mouth and my eyes roll back in my head. "Dead." "Oh!" he yelled. "*Friedhof!*" Yes! *Friedhof!* How had I let myself be so lousy a tourist as to leave the apartment without the word *Friedhof*? We almost jumped up and down together. "*Friedhof!* Yes, yes!" He pointed up the hill and around the corner. We shook hands warmly, and I walked on. Eventually I found myself skirting an endless concrete wall. The highway was still wide and busy, but buildings seemed to be farther apart, and perhaps there was a residential district ahead in the area that seemed to have sun rather than gray hovering over it. But I saw no graveyard. When a young woman walked by pushing her fancy German baby stroller, I asked, "St. Marx Friedhof?" and in precise, heavily accented English, she told me that it was just ahead, perhaps ten more minutes. "Are you searching for Mozart?" "Um, yes, actually." "He is just up the central road in the cemetery grounds, turn left at the iron cross." I thanked her and wondered if, at any moment, I would be joining a stream of Mozart pilgrims, all of us lining up to turn left at the cross. But no. I finally found the tall brick archway, its iron gate ajar. I walked through. A graveyard quiet descended, and I was in another world, completely alone.

I had heard that the cemetery was reclaimed and restored in the 1970s, so I was expecting a pretty, parkish place. Instead, there was something much better. The upkeep appeared to involve just a quick and occasional mowing around the central road, as the woman had called it—more of a wide, stone-strewn path. Beyond this were acres of graves, all of them a hundred and fifty years old, two hundred, more. They were surrounded by ancient tangled pines, chestnuts, maples. Weedy grasses grew among the rows and rows of grave markers, all green with moss and lichens. There were angel heads and devil wings and statues of little girls gazing up into the face of God. All of them were losing heads and limbs; the words on the markers were often too worn, too crumbled, to read. There were demons and gargoyles and the hushed whispers of spirits. There were also birds—chickadees, and English robins, and arguing crows. (Vienna is very urban; there are not many trees. Though I dutifully carried my little binoculars everywhere, I had seen very few birds—suddenly I was surrounded.) To my mind, this cemetery was the quietest, most magical, most beautiful, most haunted place in all of Vienna. Eventually I did see a few other people, graveyard wanderers, just here and there. But solitude was easy to come by, even at the memorial of the cemetery's most famous inhabitant. Just as the woman with the stroller had told me, up the hill I found a small, old, iron sign, with an arrow pointing left and the word MOZARTGRAB. Another twenty yards or so, there was the memorial. I gathered fallen chestnuts and dropped them into my pockets as I approached the circle of raked pebbles and a lackluster stone.

No one knows where Mozart is actually buried. To save space, and to allow for the turnover of graves every ten years (bones were haphazardly dug up and replaced with new bodies), Joseph mandated that there be no gravestones on the actual graves in his new cemetery. Instead, individual grave markers were lined up along the fence, often some distance from the place of burial. Seventeen years after Mozart was buried, with interest in his biography growing, Constanze and her new husband tried to locate the gravesite, but no one could tell them anything. Gravediggers waved their shovels in the general direction of the area that Mozart was rumored to lie. There is a skull held by the Mozarteum in Salzburg that early gravediggers claimed they had removed from Mozart's grave, but their story has been discredited. Modern sleuthing has revealed nothing more.

In the general vicinity of Mozart's burial place, there was once a granite memorial with a larger-than-life-size statue of a mourning Muse, but this was moved in 1874 to the Zentralfriedhof, the central municipal cemetery, to stand alongside the memorials for all the other great composers buried in Vienna—Beethoven, Schubert, Salieri, Brahms, Strauss, Schoenberg, and many more. A ghostly garden of musicians. The statue is beautiful, and the cemetery is well kept, but it is far busier with tourists than St. Marx. I much preferred the haunted, gothic silence of the smaller burial ground.

Now the gray had lifted, and the day was all sun and birds. But the preternatural stillness of the graveyard was palpable, simultaneously beautiful and eerie. I walked over to Mozart's memorial and sat down at the base of the stone. There was a pillar inlaid with a marble plate that said, simply, W. A. MOZART 1756–1791. Leaning on the base of the pillar was a pale carved cherub, perhaps half my height. His waist was wrapped in a loose cloth, and he held the end of it in one hand, maybe to keep it from falling off. His other elbow leaned on the pillar, his head in his hand. I crouched to look up at his face, expecting it to capture a sense of loss, of sadness, or perhaps a wistful listening to the music of the heavens. But this cherub conveyed none of these things. His attitude was completely disaffected, aloof, almost annoyed, as if to say, *Seriously? I have to stand here forever and it's not even his grave?* The cherub's feet were covered, this day, in blooming pink begonias and a recent offering of cut yellow roses, just beginning to wilt. I laughed at the plight of the cherub.

And yet Mozart has been honored at this exact place for more than two hundred years. Even though the original memorial was moved, true Mozart devotees make their way here, to St. Marx. Thousands upon thousands have visited; at this very spot, there have been millions of prayers and dreams of music and personal wishes and raising of faces to heaven in search of Mozart's spirit. This graveyard does in fact somewhere cradle the dust of the maestro's bones. And yes, I felt something. A spirit, a rush, a presence. A sadness. A consolation. The breeze lifted my hair and set it back down again. My imagination ran happy and wild.

The disgruntled cherub at Mozart's memorial, St. Marx Cemetery, Vienna. *(Photograph by the author)*

By the time Star died, the Mozarts had been forced by financial constraints to leave their beloved Domgasse rooms and move to smaller apartments outside the town center. Their new lodging was on Landstrasse, not far from St. Marx Cemetery, where Mozart would be buried. While planning my journey to Vienna I dreamed of a little pilgrimage I would make, walking somber and peaceful and wistful, from the graveyard to the site of these lodgings. Here I would sneak about the grounds, or if the current owner was home and seemed kind, I would ask whether I might walk in the garden. I was sure that after all my thinking and imagining about Star's funeral, I would somehow intuit which tiny patch of garden was the likely gravesite of Mozart's starling. So I had to laugh at my disillusionment when I discovered that the place where I'd originally disembarked the tram—the industrial area with its giant buildings decorated with the logo of a prominent mobile phone company—was just the place the Mozarts had lived when Star died. No old houses. Not even new houses. Not a shred of earth. Just concrete and cars and a wasteland of industry as far as the eye could see.

Even so. It was silly. It was a flight of fancy, I knew. But the chestnut trees in the graveyard were so old, so twisted, so lovely. The autumn chestnuts on the ground around Mozart's memorial were so round, so glowing brown. Mozart had been buried in this earth for ten years or so before what was left of his bones had been raked aside to make space for new bodies. Mozart's body could have nourished these trees. These chestnuts in my pocket, to my mystical mind, bore an authentic connection to his physical presence on earth, and the fruitfulness of his work beyond that life. It is not such a strange notion—a material continuance, a sense of life after death that is both poetically mystical and fully earthen. Rachel Carson articulated this sensibility in her article "Undersea," published in a 1937 issue of the *Atlantic Monthly*: "Individual elements are lost to view, only to reappear again and again in different incarnations in a kind of material immortality... Against this cosmic background the life span of a particular plant or animal appears not as a drama complete in itself, but only as a brief interlude in a panorama of endless change."

Clutching my Mozart-imbued chestnuts, I asked Siri to help me find the exact address of the Mozarts' Landstrasse apartment. I stood on the sidewalk, scanned the wires, and hoped starlings would appear. They would provide psychological closure, and a tidy ending for this book. But none did. I thought I might at least enjoy a quiet moment of contemplation here on the concrete, but now my tram, the 71 death tram, was fast approaching. There wouldn't be another for more than half an hour and I was hungry. Quickly, rashly, joyfully, I pulled the roundest and fattest chestnut from my pocket, kissed it, and tucked it into a corner of the sidewalk. Star's new grave marker, as accurate, at least, as Mozart's. With a strange energy, I jumped on the bus.

While I wandered the dreamy quiet of St. Marx Friedhof, it was the *Requiem* that swirled through my head. But when I set my chestnut on the gray concrete that had to stand in for Star's tiny, forgotten grave, it was the wild, swirling cadenzas from *A Musical Joke* that filled my mind and heart. Even more than his poem, this flight of musical fancy was Mozart's truest elegy for his small friend, the commonest of birds who could never have known that he was joining with a musical genius in the highest purpose of creative life: to disturb us out of complacency; to show us the wild, imperfect, murmuring harmony of the world we inhabit; to draw our own lives into the song.

Coda

One of the queries I get frequently from friends is what I will "do with Carmen" now that I've finished this book. The answer seems obvious to me. Carmen is a member of my household, and I will look after her as long as she lives. There is no telling how long that will be. A good age for a wild starling is six years or so, but captive birds can die young from some wisp of a draft or live for fifteen years and more. If for some reason I arrive at a place in my life where I cannot keep her, then I will find a home where she will be welcomed with delight. But I'll offer one last thought: Carmen brings joy and depth and insight to our family. I believe she has a good life, and I am glad she did not die with her nest mates. But not one single day passes that I do not wish I could see her fly free.

ACKNOWLEDGMENTS

More than anything else I've written, *Mozart's Starling* was created in community, beneath a "cloud of witnesses." I am profoundly grateful to all who nourished this project in so many ways. For the sharing of insight and expertise: the staff at the Stiftung Mozarteum in Salzburg, especially Dr. Ulrich Leisinger; the staff at Mozarthaus in Vienna, particularly Constanze Hell; Speight Jenkins, Seattle Opera director emeritus; Dave Beck of Classical KING FM; Dr. Adela Ramos, Pacific Lutheran University; Dr. Timothy Gentner, UC San Diego; Dr. Meredith West, Indiana University; Dr. David Rothenberg, New Jersey Institute of Technology; Dr. Walter Koenig, Cornell Lab of Ornithology; Dr. Dennis Paulson, University of Puget Sound Slater Museum, emeritus; David Maskowitz, master tracker; Chris Anderson, Washington Department of Fish and Wildlife; Pat Burnett, Edmonds Community College; Samantha Randall; Dr. Christopher Plumb; Dr. Albert Furtwangler; Dell Gossett; Dr. Rob Duisberg; all who contributed their knowledge of the German language; and so many others who assisted along the way. For hospitality: the ever-fabulous Benedictine sisters of St. Placid Priory; Cathy Cowell and all the good people at Whiteley Center, Friday Harbor; Wendy Dion at Yoga

Lodge, Whidbey Island; and my lovely Airbnb hosts in Vienna and Salzburg. For writerly commiseration: Maria Dolan, David Laskin, David Williams, Lang Cook, Martha Silano, Susan Tweit, Kathryn True, Anne Linnea, Michelle Goodman, Sage Cohen, Lynda Mapes, Anne Copeland, and all the Seattle7 writers. For more support than I deserve: the Unspeakables. For generous and intelligent professional guidance: my editor, Tracy Behar, at Little, Brown; my literary agent, Elizabeth Wales; and my copyeditors, Tracy Roe and Pamela Marshall. For undertaking the thankless task of long-term Carmen sitting (a favor one could only ask of dear friends): Trileigh Tucker, Mark Ahlness, Janeanne Houston, and Jane Davis. For the sharing of comfort and joy during the winding writing process, thanks to my family: cutest-ever parents, Jerry and Irene Haupt; amazing sister, Kelly Haupt, and sister-in-law, Jill Story; and sweet (because I know they love being called "sweet") parents-in-law, Ginny and Al Furtwangler. For creative mayhem: Carmen; Delilah; hens Ophelia, Ethel, Winifred, and Pansy (rest in peace, Marigold); and all the more-than-human creatures that crossed my path during the writing of this book. For intrepid spirit: the memory of Idie Ulsh (1936–2015), naturalist-doyenne of the Pacific Northwest, who inspired thousands with her delight in the wild world (Idie, I still want to be you when I grow up). For courage: Rachel Carson, Beatrix Potter, Georgia O'Keeffe, and all the poets, writers, artists, and composers who found their creativity upon the wild earth and whose work visits across time in the form of word, line, song, and sound. And for love: Tom and Claire.

BIBLIOGRAPHY

Adret-Hausberger, Martine. "Social Influences on the Whistled Songs of Starlings." *Behavioral Ecology and Sociobiology* 1, no. 4 (1982): 241–46.

Anderson, Stephen R. "The Logical Structure of Linguistic Theory." *Language* 84, no. 4 (2008): 795–814.

Araya-Salas, M. "Is Birdsong Music? Evaluating Harmonic Intervals in Songs of a Neotropical Songbird." *Animal Behaviour* 84, no. 2 (2012): 309–13.

Armstrong, Edward Allworthy. *A Study of Bird Song.* London: Oxford University Press, 1963.

Baserga, R. "Bird Song in His Heart." *Science* 291, no. 5510 (2001): 1902, doi: 10.1126/science.291.5510.1902a.

Bates, B. K. "Genes Tell Story of Birdsong and Human Speech." http://today.duke.edu/2014/12/vocalbird.

Bent, Arthur Cleveland. *Life Histories of North American Wagtails, Shrikes, Vireos, and Their Allies.* New York: Dover Publications, 1965.

Berger, Jonathan. "How Music Hijacks Our Perception of Time." *Nautilus,* January 23, 2014. http://nautil.us/issue/9/time/how-music-hijacks-our-perception-of-time.

Bertin, Aline, Martine Hausberger, Laurence Henry, and Marie-Annick Richard-Yris. "Adult and Peer Influences on Starling Song Development." *Developmental Psychobiology* 49, no. 4 (2007): 362–74.

Bower, Bruce. "The Pirahã Challenge." *Science News* 168, no. 24 (2005): 376–80.

Braunbehrens, Volkmar. *Mozart in Vienna: 1781–1791.* London: Deutsch, 1990.

Breittruck, Julia. "Pet Birds. Cages and Practices of Domestication in Eighteenth-Century Paris." *InterDisciplines: Journal of History and Sociology* 3, no. 1 (2012): 1–48.

Cabe, P. R. *European Starling: Sturnus vulgaris.* Washington, DC: American Ornithologists' Union, 1993.

Carlic, Steve. "Introducing America's Most Hated Bird: The Starling." Syracuse.com, September 7, 2009.

Carson, Rachel. "How About Citizenship Papers for the Starling?" *Nature* 32, no. 6 (1959): 317–19.

Chaiken, Marthaleah, and Jörg Böhner. "Song Learning After Isolation in the Open-Ended Learner the European Starling: Dissociation of Imitation and Syntactic Development." *Condor* 109, no. 4 (2007): 968–76.

Chaiken, Marthaleah, Jörg Böhner, and Peter Marler. "Song Acquisition in European Starlings, *Sturnus vulgaris:* A Comparison of the Songs of Live-Tutored, Tape-Tutored, Untutored, and Wild-Caught Males." *Animal Behaviour* 46, no. 6 (1993): 1079–90.

Christie, Douglas E. *The Blue Sapphire of the Mind: Notes for a Contemplative Ecology.* New York: Oxford University Press, 2013.

Colapinto, John. "The Interpreter: Has a Remote Amazonian Tribe Upended Our Understanding of Language?" *New Yorker,* April 16, 2007.

Comins, Jordan A., and Timothy Q. Gentner. "Auditory Temporal Pattern Learning by Songbirds Using Maximal Stimulus Diversity and Minimal Repetition." *Animal Cognition* 17, no. 5 (2014): 1023–30.

———. "Pattern-Induced Covert Category Learning in Songbirds." *Current Biology* 25, no. 14 (2015): 1873–77.

———. "Perceptual Categories Enable Pattern Generalization in Songbirds." *Cognition* 128, no. 2 (2013): 113–18.

———. "Working Memory for Patterned Sequences of Auditory Objects in a Songbird." *Cognition* 117, no. 1 (2010): 38–53.

Corbo, Margarete Sigle, and Diane Marie Barras. *Arnie, the Darling Starling.* Boston: Houghton Mifflin, 1983.

Costanza, Stephen. *Mozart Finds a Melody.* New York: Henry Holt, 2004.

Davenport, Marcia. *Mozart.* New York: C. Scribner's Sons, 1932.

Deutsch, Otto Erich. *Mozart: A Documentary Biography.* Stanford: Stanford University Press, 1965.

de Waal, Frans. *The Ape and the Sushi Master: Cultural Reflections of a Primatologist.* New York: Basic Books, 2001.

Doolittle, Emily, and Henrik Brumm. "O Canto do Uirapuru: Consonant Intervals and Patterns in the Song of the Musician Wren." *Journal of Interdisciplinary Music Studies* 6, no. 1 (2012): 55–85.

Eens, Marcel, Rianne Pinxten, and Rudolf Frans Verheyen. "Male Song as a Cue for Mate Choice in the European Starling." *Behaviour* 116, no. 3 (1991): 210–38.

Estés, Clarissa Pinkola. *Women Who Run with the Wolves: Myths and Stories of the Wild Woman Archetype*. New York: Ballantine, 1992.

Everett, Daniel. "Chomsky, Wolfe, and Me." *Aeon*, January 10, 2017.

Gentner, Timothy Q. "Mechanisms of Temporal Auditory Pattern Recognition in Songbirds." *Language Learning and Development* 3, no. 2 (2007): 157–78.

Gentner, Timothy Q., Kimberly M. Fenn, Daniel Margoliash, and Howard C. Nusbaum. "Recursive Syntactic Pattern Learning by Songbirds." *Nature* 440, no. 7088 (2006): 1204–7.

Glover, Jane. *Mozart's Women: The Man, the Music, and the Loves of His Life*. New York: HarperCollins, 2005.

Godman, Stanley, and Stanley Godwin. *The Bird Fancyer's Delight*. London: Schott, 1985.

Gurewitsch, Matthew. "An Audubon in Sound." *Atlantic* 279, no. 3 (1997): 90–96.

Hartshorne, C. *Born to Sing: An Interpretation and World Survey of Bird Song*. Bloomington: Indiana University Press, 1973.

Haupt, Lyanda Lynn. *The Urban Bestiary: Encountering the Everyday Wild*. Boston: Little, Brown, 2012.

Hausberger, M. "Organization of Whistled Song Sequences in the European Starling." *Bird Behavior* 9, no. 1 (1990): 81–87.

Hausberger, M., P. Jenkins, and J. Keene. "Species-Specificity and Mimicry in Bird Song: Are They Paradoxes? A Reevaluation of Song Mimicry in the European Starling." *Behaviour* 117, no. 1 (1991): 53–81.

Hauser, Marc D., Noam Chomsky, and W. Tecumseh Fitch. "The Faculty of Language: What Is It, Who Has It, and How Did It Evolve?" *Science* 298, no. 5598 (2002): 1569–79.

Healy, Kevin, Luke McNally, Graeme D. Ruxton, Natalie Cooper, and Andrew L. Jackson. "Metabolic Rate and Body Size Are Linked with Perception of Temporal Information." *Animal Behaviour* 86, no. 4 (2013): 685–96.

Heartz, Daniel. "Mozart's Sense for Nature." *Nineteenth-Century Music* 15, no. 2 (1991): 107–15.

Hewett, I. "What Humans Can Learn from Birdsong." http://www.telegraph.co.uk/culture/music/classicalmusic/10841468/What-humans-can-learn-from-birdsong.html.

Hindmarsh, Andrew M. "Vocal Mimicry in Starlings." *Behaviour* 90, no. 4 (1984): 302–24.

Hockett, Charles. "Animal 'Languages' and Human Language." *Human Biology* 31, no. 1 (1959): 32–39.

Hulse, Stewart H., John Humpal, and Jeffrey Cynx. "Discrimination and Generalization of Rhythmic and Arrhythmic Sound Patterns by European Starlings (*Sturnus vulgaris*)." *Music Perception: An Interdisciplinary Journal* 1, no. 4 (1984): 442–64.

Humphreys, Rob. *The Rough Guide to Vienna.* London: Rough Guides, 2011.

Hutchings, Arthur. *A Companion to Mozart's Piano Concertos*. London: Oxford University Press, 1950.

Jarvis, Erich D. "Brains and Birdsong." *Nature's Music* (2004): 226–71.

———. "Learned Birdsong and the Neurobiology of Human Language." *Annals of the New York Academy of Sciences* 1016, no. 1 (2004): 749–77.

Jenkins, Speight. "Papageno's Magical Humanity." *Encore* (May 2011): 10–11.

Johnson, P. *Mozart: A Life*. New York: Viking, 2013.

Keefe, S. P. *The Cambridge Companion to Mozart*. Cambridge: Cambridge University Press, 2003.

Koenig, W. D. "European Starlings and Their Effect on Native Cavity-Nesting Birds." *Conservation Biology* 17, no. 4 (2003): 1134–40.

Krause, Bernard L. *The Great Animal Orchestra: Finding the Origins of Music in the World's Wild Places*. Boston: Little, Brown, 2012.

Landon, H. C. Robbins. *Mozart and Vienna*. New York: Schirmer, 1991.

Levi, Erik. *Mozart and the Nazis: How the Third Reich Abused a Cultural Icon*. New Haven, CT: Yale University Press, 2010.

Lorenz, Michael. "New and Old Documents Concerning Mozart's Pupils Barbara Ployer and Josepha Auernhammer." *Eighteenth-Century Music* 3, no. 2 (2006): 311–22.

Lowe, Steven. "Mozart Piano Concertos Nos. 17 & 24." *Seattle Symphony Program Notes* (May 2015): 26–27.

Marcus, Gary F. "Language: Startling Starlings." *Nature* 440, no. 7088 (2006): 1117–18.

Mannes, Elena. *The Power of Music: Pioneering Discoveries in the New Science of Song.* New York: Walker, 2011.

McClary, Susan. "A Musical Dialectic from the Enlightenment: Mozart's Piano Concerto in G Major, K. 453, Movement 2." *Cultural Critique,* no. 4 (1986): 129–69.

Melograni, P., and L. G. Cochrane. *Wolfgang Amadeus Mozart: A Biography.* Chicago: University of Chicago Press, 2007.

Milius, Susan. "Grammar's for the Birds: Human-Only Language Rule? Tell Starlings." *Science News,* April 26, 2006. https://www.sciencenews.org/article/grammars-birds-human-only-language-rule-tell-starlings.

———. "Music Without Borders." *Science News* 157, no. 16 (2000): 252.

Mozart, W. A. *Mozart: A Life in Letters.* Edited by Cliff Eisen. Translated by Stewart Spencer. New York: Penguin, 2006.

———. *Mozart's Letters, Mozart's Life: Selected Letters.* Edited and translated by Robert Spaethling. New York: Norton, 2000.

O'Donohue, J. *Anam Cara: A Book of Celtic Wisdom.* New York: Cliff Street Books, 1997.

Osborne, Charles. *The Complete Operas of Mozart: A Critical Guide.* New York: Atheneum, 1978.

Parks, G. Hapgood. "A Convenient Method of Sexing and Aging the Starling." *Bird-Banding* 33, no. 3 (1962): 148.

Pavlova, Denitza, Rianne Pinxten, and Marcel Eens. "Female Song in European Starlings: Sex Differences, Complexity, and Composition." *Condor* 107, no. 3 (2005): 559–69.

Plumb, Christopher. *The Georgian Menagerie: Exotic Animals in Eighteenth-Century London*. London: I. B. Tauris, 2015.

Robbins, Louise E. *Elephant Slaves and Pampered Parrots: Exotic Animals in Eighteenth-Century Paris*. Baltimore: Johns Hopkins University Press, 2002.

Rosen, Charles. *The Classical Style: Haydn, Mozart, Beethoven*. New York: W. W. Norton, 1997.

Rothenberg, David. *Why Birds Sing: A Journey Through the Mystery of Bird Song*. New York: Basic Books, 2005.

Sadie, Stanley. *The New Grove Mozart*. New York: W. W. Norton, 1983.

Shaffer, Peter. *Amadeus: A Drama*. New York: Samuel French, 2003.

Sibley, David, Chris Elphick, and John B. Dunning. *The Sibley Guide to Bird Life and Behavior*. New York: Alfred A. Knopf, 2001.

Solomon, Maynard. *Mozart: A Life*. New York: HarperCollins, 1995.

Stalzer, Alfred. *Mozarthaus Vienna*. Munich: Prestel, 2006.

Steves, Rick. *Vienna, Salzburg, and Tirol*. Berkeley, CA: Avalon Travel Publishing, 2014.

Sund, Patricia. "History of Pet Bird Cages." http://www.birdchannel.com/bird-housing/bird-cages/history-of-cages.aspx.

Thomas, Keith. *Man and the Natural World: Changing Attitudes in England, 1500–1800*. London: Allen Lane, 1983.

Till, N. *Mozart and the Enlightenment: Truth, Virtue, and Beauty in Mozart's Operas.* New York: W. W. Norton, 1993.

Tinbergen, J. M. "Foraging Decisions in Starlings (*Sturnus vulgaris* L.)." *Ardea* 69 (2002): 1–67.

Todd, Kim. *Tinkering with Eden: A Natural History of Exotics in America.* New York: W. W. Norton, 2001.

Todt, Dietmar. "Influence of Auditory Stimulation on the Development of Syntactical and Temporal Features in European Starling Song." *Auk* 113, no. 2 (1996): 450–56.

Van Heijningen, Caroline A. A., Jos de Visser, Willem Zuidema, and Carel ten Cate. "Simple Rules Can Explain Discrimination of Putative Recursive Syntactic Structures by Songbirds: A Case Study on Zebra Finches." *The Evolution of Language.* Singapore: World Scientific Publishing, 2010.

Waltham, David. *Lucky Planet: Why Earth Is Exceptional—and What That Means for Life in the Universe.* New York: Basic Books, 2014.

Wayman, E. "Speech, Birdsong Share Wetware: Similar Genes Process Human and Bird Communication." *Science News* 183, no. 5 (2013).

Welty, Joel Carl, and Luis Felipe Baptista. *The Life of Birds.* New York: Saunders, 1988.

West, Meredith J., and Andrew P. King. "Mozart's Starling." *American Scientist* 78 (1990): 106–14.

West, Meredith J., A. Neil Stroud, and Andrew P. King. "Mimicry of the Human Voice by European Starlings:

The Role of Social Interaction." *Wilson Bulletin* 95, no. 4 (1983): 635–40.

Wydoski, Richard S. "Seasonal Changes in the Color of Starling Bills." *Auk* 81, no. 4 (1964): 542–50.

Zaslaw, N., and W. Cowdery, eds. *The Compleat Mozart: A Guide to the Musical Works of Wolfgang Amadeus Mozart.* New York: W. W. Norton, 1990.

ABOUT THE AUTHOR

Lyanda Lynn Haupt is an ecophilosopher, naturalist, and author of several books, including *The Urban Bestiary, Crow Planet, Pilgrim on the Great Bird Continent,* and *Rare Encounters with Ordinary Birds.* A winner of the Washington State Book Award and the Sigurd F. Olson Nature Writing Award, she lives in Seattle with her husband and daughter.